International Research in
the Antarctic

International Research in the Antarctic

RICHARD FIFIELD

Published for the
Scientific Committee on Antarctic Research
(SCAR) and the ICSU Press by
Oxford University Press
1987

Oxford University Press, Walton Street, Oxford OX2 6DP

Oxford New York Toronto
Delhi Bombay Calcutta Madras Karachi
Petaling Jaya Singapore Hong Kong Tokyo
Nairobi Dar es Salaam Cape Town
Melbourne Auckland

and associated companies in
Beirut Berlin Ibadan Nicosia

Oxford is a trade mark of Oxford University Press

Published in the United States
by Oxford University Press, New York

British Library Cataloguing in Publication Data
Fifield, Richard
International research in the Antarctic
1. Antarctic regions
I. Title   II. Scientific Committee on Antarctic Research
998'.9   G860
ISBN 0–19–854216–X

Library of Congress Cataloging in Publication Data
Fifield, Richard
International research in the Antarctic
Bibliography: p.
Includes index
1. Research—Antarctic regions.   2. International
Council of Scientific Unions. Scientific Committee on
Antarctic Research.   I. Title
Q180.A6F54 1987   507'.20989   87–12371
ISBN 0–19–854216–X

Typeset by Butler and Tanner Ltd, Frome, Somerset
Printed and bound in Hong Kong

PREFACE

NEW BOUNDS OF INTERNATIONAL
CO-OPERATION IN SCIENCE

I find myself in a dilemma. Although every book has to have an author or an editor, and my name appears on this one, I am, in fact, neither the author nor the editor of it. I have written about the science of the Antarctic; however, I have not studied Antarctic sciences *per se*, nor have I travelled that far South! My role in the production of this book is really to conduct for as wide an audience as is possible an account of the activities, of the Scientific Committee on Antarctic Research—familiar to all who know of it as SCAR. I am, then, the 'conductor' of this opus.

I believe that the way that SCAR, the International Council of Scientific Unions (ICSU), and I have chosen to treat our subject is not an unreasonable one. The Antarctic is proof that great things can be achieved through co-operation. What has been achieved in the name of science in the Antarctic region has been, and is being, achieved largely through international co-operation, between scientists and scientists, and governments and governments.

The Antarctic region, as this book shows, presents a macrocosm of the sciences, atmospheric, geologic, oceanographic, biologic, human, and environmental. The continent and the seas provide for science vast natural laboratories for learning, experimenting, detecting, testing, finding, observing, and measuring. Here is an unrivalled access to the Earth's magnetic field, our upper atmosphere and out into space. Here are large ecosystems and food chains. Here for the investigating are almost unfathomable depths of frozen ice that contains archives of frozen atmospheres millions of years old. And these archives are not just relevant to studies of the past but also of the present and indeed the future of our planet Earth. The ice and the life that inhabits the edge of the continent and the surrounding seas, provide sensitive indicators of mankind's activities, and especially his industry, and energy and transport systems in the continents to the north. The

Antarctic region also plays a vital part in stabilizing the climate of our planet and in establishing a flow of deep cold bottom waters into the surrounding oceans.

The work that I was asked to conduct was essentially to bring together in a readily understandable form an account of the achievements of Antarctic science, of the Working Groups and Specialist Committees of SCAR, and their involvements with other international bodies of scientists and scientific institutions and to describe the historical background to the present-day extensive scientific interest in this remote part of our planet.

I must express my appreciation to the SCAR and ICSU Secretariats, to the chief officers of the many international groups of SCAR and those of its sister organization, the Scientific Committee for Oceanic Research, who provided me with material and first drafts of the chapters covering their specialities and the pictures to illustrate them. Also I am grateful to SCAR for facilitating my access to scientists active in Antarctic studies and to their home-based laboratories, some in the Northern Hemisphere and some in the Southern Hemisphere. Sometimes I have been able to lead them, but often they have lead me and brought me back on to a sensible path. To be master of all the sciences that they collectively represent is not possible. It is their science that I present in this book. It is their wish that the pictures and threads of science that emerge be seen to come largely from the spirit of cooperation. It is those many but unnamed individuals active in the affairs of SCAR who have made this book possible.

I owe particular thanks to many colleagues at *New Scientist*, to Michael Kenward, the journal's Editor, for his forebearance in the time that it has taken to complete this task, to Karen Iddon, Mandy Caplan, and Jane Moore for typing, and retyping, the manuscripts, to Neil Hyslop for his painstaking work on the many diagrams, to Cathy

Moore for her advice and help in matters geographical, to Dr John Gribbin and Dr Christine Sutton for their assistance with some of the work on the upper atmosphere, and to Dr Stephanie Pain for helpful advice on the biology. Last but not least, a special thank you to the staff of Oxford University Press.

London   Richard Fifield
March 1987

CONTENTS

The demand of science, that no part of the globe shall remain untouched by the hand of investigation, was the force that drew our little band to the land of the farthest south.

OTTO NORDENSKJOLD
Antarctica (Hurst and Blackett, London, 1905.)

INTRODUCTION

THE SCIENTIFIC COMMITTEE ON ANTARCTIC RESEARCH*

The aim of this book is to present to the public the results of scientific research in the Antarctic, in the belief that the broader the base of understanding about mankind's quest for knowledge in this far-off land, the more likely it will be that the nations of the world will continue their support of future research in the southern polar regions of the planet Earth.

The book is the outgrowth of an idea that an account of the scientific aspects of the Antarctic, written for the non-scientist, would be a useful addition to the literature. The initiator of the book is the Scientific Committee on Antarctic Research, commonly referred to by its acronym, SCAR. SCAR is a non-governmental body organized under its parent, the International Council of Scientific Unions (ICSU), and has been in existence since the International Geophysical Year (IGY) of 1957–8. SCAR is the only organization devoted solely to the promotion and co-ordination of research in several scientific disciplines as they relate to the understanding of natural phenomena in the Antarctic.

SCAR AND THE ANTARCTIC TREATY

No discussion of Antarctic science would be complete without reference to SCAR and the Antarctic Treaty. The Treaty provides for free and open access to the Antarctic by all nations engaged in ongoing scientific studies there. SCAR is not a part of the Treaty, but promotes and fosters international co-operation in the pursuit of Antarctic science.

SCAR and the Treaty have functioned independently side by side since both came into existence after the IGY. Without their mutually supportive relationship, it is doubtful that Antarctic science would have reached its present level of sophistication. The continuation of SCAR and the Treaty are fundamental to the continuation of further scientific discoveries in the Antarctic.

The Treaty operates in the world of international diplomacy, while SCAR operates in the world of international science. The Treaty functions through the workings of governmental representatives, while SCAR functions through the workings of scientists, who act independently of their governments. The Treaty is expressed in terms of a precise document signed and ratified by the member nations, while SCAR is expressed in terms of a constitution agreed on by its member delegates, subject to the approval of ICSU. Despite these differences, the Treaty and SCAR have had a mutually beneficial relationship over the years, and it is difficult to conceive of a sound policy for the Antarctic without both organizations acting independently on the one hand, but with interacting roles on the other.

ORIGIN AND GROWTH OF SCAR

The beginnings of SCAR. The IGY began on 1 January 1957, and ended officially on 31 December 1958. Even before the IGY began, however, it was recognized that 18 months was too short in terms of the investment in stations and equipment that was being made by the nations planning extensive activities in the Antarctic. Following a proposal made by the US National Committee in 1956 to the Comité Special de l'Année Geophysique Internationale (CSAGI), the IGY programmes were extended for an additional year in order to justify the huge expenditures that were being made for Antarctic research, and ICSU established SCAR in 1958 (see Chapter 3). SCAR was thus born of the desire of scientists to continue the international co-ordination of research in the Antarctic following the IGY.

* Based on 'The Antarctic Treaty as a Scientific Mechanism—The Scientific Committee on Antarctic Research and the Antarctic Treaty System', by James H. Zumberge, Chapter 10 in *The Antarctic Treaty System: an Assessment* (National Academy of Sciences, Washington DC 1986.)

Early years of SCAR. The first meeting of SCAR was held in The Hague in February 1958. The main order of business was the drafting of a constitution, the election of officers, and the preparation of a budget and making provisions for funding it. The SCAR constitution was ratified by ICSU at its 8th general assembly in October 1958. The first officers of SCAR were also elected at The Hague.*

The SCAR budget was set at $6000 per year, with each of the 12 members contributing $500 toward this amount. To cover larger budgets in future years, it was decided that members would contribute additional amounts in proportion to the level of Antarctic activity as measured by the number of wintering-over personnel. This criterion was dropped in 1980, however, because of the growth of important summer programmes including extensive marine science studies.

Growth in SCAR membership. The membership of SCAR remained constant for 20 years until Poland and the Federal Republic of Germany were admitted in 1978, bringing the total representation to 14 countries. Before then, however, representation from other organizations under the ICSU umbrella had increased somewhat.

The fifteenth country to gain representation in SCAR was the German Democratic Republic, whose delegate was admited in 1981. The sixteenth and seventeenth representatives, from Brazil and India, were voted into SCAR in 1984. Finally, the People's Republic of China gained admission at the XIX SCAR meeting in 1986. Table 1 gives a list of all SCAR member nations with national Antarctic committees and the years in which they were admitted to SCAR.

SCAR STRUCTURE AND PROCEDURES

The SCAR constitution. The SCAR constitution was a rather short and simple document when first formulated at The Hague in 1958. It consisted of a preamble, criteria for membership, and the basic principles to guide SCAR's functioning.

The preamble stated that 'SCAR is a Special Committee of ICSU charged with furthering the co-ordination of scientific activity in the Antarctic, with a view to framing a scientific programme of

* SCAR Presidents 1958–86: G. R. Laclavère (France), 1958–63; L. M. Gould (USA), 1963–70; G. de Q. Robin (UK), 1970–4; T. Gjelsvik (Norway), 1974–8; G. A. Knox (NZ), 1978–82; J. H. Zumberge (USA), 1982–6; C. Lorius (France) 1986–90.

circumpolar scope and significance. In establishing its programme, SCAR will take care to acknowledge the autonomy of other existing international bodies.'

The constitution defined the membership as one delegate from 'each country actively engaged in

Table 1. Nations with national Antarctic committees that are members of SCAR[a]

Country	Year admitted
Argentina	1958
Australia	1958
Belgium	1958
Brazil	1984
Chile	1958
China, People's Republic	1986
France	1958
German Democratic Republic	1981
Germany, Federal Republic of	1978
India	1984
Japan	1958
New Zealand	1958
Norway	1958
Poland	1978
South Africa	1958
Union of Soviet Socialist Republics	1958
United Kingdom	1958
United States of America	1958
Associate members:	1987
Spain, the Netherlands, Peru, Sweden[b], Switzerland[c]	

[a] As of April 1987. [b] As of May 1987. [c] As of June 1987.

Antarctic research' plus one delegate from each of the international scientific unions federated in ICSU which desire to participate in SCAR. Other special committees of ICSU could send observers to SCAR meetings.

The balance of the constitution dealt with the establishment of the SCAR executive (president, vice president, and secretary), authority to establish *ad hoc* committees, and a procedure for preparing the budget and fixing contributions by members as recommended by the budget committee.

The constitution has been revised slightly from time to time and the present version includes the following statement, under the section entitled 'Guidelines for the Conduct of SCAR Affairs': 'SCAR will abstain from involvement in political and juridical matters, including the formulation of management measures for exploitable resources, except where SCAR accepts an invitation to advise on a problem'.

The present constitution also provides for an

alternate delegate in addition to the permanent delegate from each national committee adhering to SCAR, but each country is entitled to only one vote. The consitution also details an expanded SCAR executive consisting of a president, immediate past-president, two vice presidents, and a secretary.

The general condition for national membership remains unchanged; that is, only countries actively engaged in Antarctic research are eligible. However, SCAR's rules of procedure do permit the granting of observer status to countries that are planning to establish scientific research activities in the Antarctic, and to other international scientific organizations.

The way in which SCAR conducts its business has become more sophisticated since the 1960s, but, essentially, SCAR has remained true to the principles and philosophy that have guided its activities since its inception.

SCAR executive. Continuity of leadership in SCAR is lodged in the Executive Committee of five members, each of whom holds office for four years; however, the terms are staggered to allow for continuity of leadership. The executive meets in odd-numbered years, usually, but not always, at the SCAR headquarters in the Scott Polar Research Institute, Cambridge, UK. These meetings are designed to maintain continuity between the biennial meetings of SCAR, and normally they consist of reviewing matters that were referred to SCAR during the previous meeting, or of considering other items that need to be acted upon before the next SCAR meeting. The SCAR executive cannot act on an application for membership because that decision rests only with the SCAR delegates at a regular meeting. The executive does review applications for membership to see that all requirements have been met, after which a recommendation is forwarded to the delegates for action at the next meeting.

Working groups. The core of SCAR lies in its eight permanent working groups: Biology, Geodesy and Cartography, Geology, Glaciology, Human Biology and Medicine, Logistics, Solid Earth Geophysics, and Upper Atmosphere Physics. Until recently, two other working groups existed. The activities of the oceanography working group were more or less merged with another ICSU committee, the Scientific Committee for Oceanic Research (SCOR). The routine aspects of synoptic meteorology now come under the aegis of the WMO working group on Antarctic Meteorology, and the International Commission for Polar Meteorology handles the scientific aspects. Each national committee has the right to appoint one member to each working group. Subject to the approval of the Executive Committee, other members may be admitted from any country, if it is believed that their knowledge is useful to the deliberations of the working group. Working groups meet during SCAR meetings or at times when it is convenient for a group to assemble in connection with some other international meeting.

Groups of specialists. When matters arise in SCAR that do not fit neatly within the purview of a single working group, groups of specialists are formed. These bodies are usually formed when multidisciplinary problems or programmes are involved, or when SCAR is requested to provide advice, for example, to the Antarctic Treaty governments. Members in groups of specialists are appointed by the Executive Committee and are not regarded as representatives of national committees. When their assignments are completed, groups of specialists are discontinued.

Currently, SCAR has six groups of specialists: Antarctic Climate Research, Antarctic Sea Ice, Seals, Southern Ocean Ecology, Structure and Evolution of the Antarctic Lithosphere, and Evolution of Cenozoic Palaeoenvironments of High Southern Latitudes.

It is worth citing an example of how a group of specialists can influence the course of Antarctic research. The Group of Specialists on the Living Resources of the Southern Ocean (later co-sponsored by the Scientific Committee for Oceanic Research as its Working Group 54) designed an extensive programme of biological research in the Southern Ocean entitled Biological Investigations of Marine Antarctic Systems and Stocks (BIOMASS). The main objective of BIOMASS was to study the ecosystem of the ocean surrounding the Antarctic continent, with special attention to the life history, distribution, and abundance of krill (*Euphausia superba*). This small shrimp-like crustacean is the main element of the food web in the Southern Ocean and is a possible candidate for human exploitation (see Chapter 12).

The value of the BIOMASS investigations will be increased once the data collected are analysed and evaluated in the BIOMASS data centre. The results of this could represent valuable information for use in managing the living resources of the Southern Ocean under the Convention for the Conservation of Antarctic Marine Living Resources of 1980.

The establishment of BIOMASS under the sponsorship of SCAR is an example of how an initiative involving multinational participation can be launched successfully. While it may not have been impossible for a single nation to put together a programme similar to BIOMASS, it seems highly improbable that such an initiative from a single country could have generated the high interest and strong enthusiasm that was encountered by the founders of BIOMASS under the sponsorship of SCAR, and the acquisition of the vast amount of data over the 10 years of the programme.

Publications of SCAR. In addition to the *SCAR Bulletin*, which is published in January, May and September of each year, in English in the *Polar Record* and in Spanish by the Instituto Antártico Argentino in Buenos Aires, SCAR publishes various reports of a special nature. These reports are *ad hoc* and do not constitute a series in any sense of the word. An example of one of these is the report on *Possible Environmental Effects of Mineral Exploration and Exploitation in Antarctica*, which was published by SCAR in 1979. Because of the importance of this subject, and because the report constituted the response of SCAR to a request from the Antarctic Treaty governments, SCAR decided that it was worth publishing. Other occasional publications by SCAR are issued from time to time (see Appendix 1).

The main vehicle for communications within the SCAR organization is the *SCAR Circular*. It provides a means to request information from national committees on various matters that SCAR wishes to address, or to convey to national committees information from the SCAR executive or the SCAR secretariat. While the *SCAR Circular*s are not archival publications in the strict sense of the word, they are serially numbered and contain much information of value and importance to the national committees.

SCAR requires each member nation, through its national committee, to submit an annual report on its ongoing programmes of research and other activities in the Antarctic. National committees are required to submit to SCAR by 30 June of each year information on research programmes of the preceding year, including the current winter season. Additionally, the national report must contain a list of the occupied stations with their latitudes and longitudes, plans for the following year for both summer and winter, and a bibliography on publications related to Antarctic research that have been published since the previous report. These national reports are distributed directly by individual SCAR national committees to other national committees so that there is a continual flow of information circulated to all who are active in Antarctic research.

SCAR takes these national reports very seriously and goes so far as to specify the format of the report and the size of the paper on which they are printed. This is the only instance in which SCAR has shown any sign of adopting bureaucratic measures, but given the use to which these reports are put, the uniformity of style and format is entirely justified. To my knowledge, no national committee has ever deviated from the form prescribed.

SCAR meetings. In the early years, SCAR met every year, but later the routine of biennial meetings became the norm. Meetings are held in one of the member countries, in response to invitations extended from national committees. The host country provides all meeting sites and other amenities to the delegates at no cost to the SCAR treasury. All delegates provide their own travel and other expenses, but SCAR may fund travel expenses for the officers.

It has been the custom of SCAR to alternate its meeting sites between member countries in the Northern and Southern Hemispheres. This informal arrangement may be difficult to follow in the future, given the fact that 12 of the 18 member countries lie north of the equator and six lie south. The dates, sites, and designated numbers of all meetings of SCAR through 1984 are given in Table 2.

Normally, meetings last for two weeks. The first week is devoted to meetings of working groups and groups of specialists, and the second week is reserved for the plenary sessions and meetings of delegates. This format has worked well and is likely to be followed in the future, even with an expanded membership.

Table 2. Dates, sites, and designated numbers of SCAR meetings, 1958–86

Meeting number[a]	City	Country	Dates
I	The Hague	The Netherlands	February 1958
II	Moscow	USSR	August 1958
III	Canberra	Australia	March 1959
IV	Cambridge	United Kingdom	August–September 1960
V	Wellington	New Zealand	October 1961
VI	Boulder	United States	August 1962
VII	Cape Town	South Africa	September 1963
VIII	Paris	France	August 1964
IX	Santiago	Chile	September 1966
X	Tokyo	Japan	June 1968
XI	Oslo	Norway	August 1970
XII	Canberra	Australia	August 1972
XIII	Jackson Hole	United States	September 1974
XIV	Mendoza	Argentina	October 1976
XV	Chamonix	France	May 1978
XVI	Queenstown	New Zealand	October 1980
XVII	Leningrad	USSR	July 1982
XVIII	Bremerhaven	Federal Republic of Germany	October 1984
XIX	San Diego	United States	June 1986

[a] SCAR refers to its meetings by Roman numerals. For example, the 1968 meeting in Tokyo is designated as X SCAR.

SCAR Secretariat. Initially, SCAR functioned only with an honorary secretary, Dr G. de Q. Robin (UK), the then director of Scott Polar Research Institute in Cambridge, from 1958 to 1970. SCAR funded part-time secretarial help from 1959 and full-time since 1963. From 1962 to 1970 administrative assistance to the Secretary was provided on a part-time basis by Mr George E. Hemmen, then on the staff of the Royal Society. He was appointed by SCAR as Executive Secretary (part-time) in 1970 when the permanent secretariat was formally established at the Scott Polar Research Institute with one full-time assistant.

INTERACTION OF SCAR WITH THE
ANTARCTIC TREATY SYSTEM

The Antarctic Treaty was signed in 1959 and entered into force on 23 June 1961, after ratification by the governments of the original 12 contracting parties, governments of the same 12 countries that constituted the initial membership of SCAR. SCAR is not explicitly mentioned in the treaty, but in the report of the first Antarctic Treaty consultative meeting, in Canberra in July 1961, SCAR was referred to several times in various recommendations. These included, among others, the following wording in Recommendation I-IV: '(1) that the free exchange of information and views among scientists participating in SCAR, and the Recommendations concerning scientific programmes and co-operation formulated by this body constitute a most valuable contribution to international scientific co-operation in Antarctica; (2) that since these activities of SCAR constitute the kind of activity contemplated in Article III of the treaty, SCAR should be encouraged to continue this advisory work which has so effectively facilitated international co-operation in scientific investigation.'

These words leave little doubt about the high regard of the Treaty parties for SCAR and its programmes. This first indication of the Treaty consultative parties' respect for SCAR has continued ever since. Whenever the Treaty parties are in need of scientific advice concerning the Antarctic, they have come to SCAR.

Requests to SCAR for advice and information are made in a formal way by the Treaty parties. These recommendations are designated by a numbering system. For example, for Recommendation VIII-14, the Roman numeral refers to the eighth Antarctic Treaty consultative meeting and the number 14 identifies a specific recommendation. Recommendations to SCAR have emanated from many of the consultative meetings and are too many in number to review in these pages. Generally, these recommendations have ranged over a variety of topics, including a resolution at the first consultative meeting urging the contracting

The Antarctic wintering stations, 1986

parties to be guided in their conservation policies by the recommendations of SCAR. Other matters on which the Treaty nations have called on SCAR for advice and guidance cover such subjects as logistics, telecommunications, living resources of the Southern Ocean, and the environmental implications of possible exploration and exploitation of mineral resources in the Antarctic. The last matter deserves some additional comment because of its general interest not only to the consultative parties, but to other nations around the world.

The question of Antarctic mineral resources had never been raised formally in SCAR until 1976. Meeting in Mendoza, Argentina, that year, XIV SCAR addressed a recommendation VIII-14 from the eighth Antarctic Treaty consultative meeting held the previous year in Oslo. That recommendation invited SCAR to 'make an assessment on the basis of available information of the possible impact on the environment of the Treaty Area and other ecosystems dependent on the Antarctic environment if mineral exploration and/or exploitation were to occur there'.

SCAR was apprehensive about its response, lest it be inferred that by taking on the assignment SCAR was tacitly endorsing a move toward exploitation of mineral resources in the Antarctic. Both SCAR and the Treaty nations had avoided the minerals issue until it was forced on their agenda by international events beyond their control. The main reason why the minerals question came to the fore at this particular time was the quadrupling of the price of crude oil in 1973–4 by the Organisation of Petroleum Exporting Countries.

To formulate its response to the Treaty nations, SCAR organized a group of specialists, which wrote a report in time for the ninth Antarctic Treaty consultative meeting, in London in October 1977. This report was also published by

SCAR in 1979. This was the beginning of an ongoing relationship between SCAR and the treaty nations on the minerals issue. SCAR's role has diminished, however, because the consultative parties have met on several occasions to forge a separate accord as part of the Antarctic Treaty system to deal with the question of the extractive industries, should they ever gain a foothold on the continent or in the continental shelves surrounding it. It should be emphasized again that SCAR's role in this matter was confined to factual information and the scientific interpretation of those facts. It must be noted, however, that many individuals connected with SCAR also serve as advisers to their respective governments on Antarctic Treaty matters. In so doing, these individuals are careful to keep their roles in SCAR separate and distinct from their roles in Treaty matters.

In summary, it can be said that the consultative parties and SCAR play separate but mutually beneficial roles in the international affairs of the Antarctic. The success of this relationship is based on two observations: First, the consultative parties derive their authority from the Antarctic Treaty. Second, the success of SCAR is based not on the authority of the SCAR constitution but rather on the experience and scientific reputations of its members and working groups. Included in these are most of the world's leading experts in Antarctic affairs, both scientific and logistic. Collectively, these experts constitute the greatest concentration of talent related to Antarctic science and attendant technology ever assembled. For this reason, the consultative parties in the Antarctic Treaty System are likely to continue their dependence on SCAR for scientific and technical information for as long as the treaty and associated agreements and conventions are in force.

SCAR'S FUTURE

During the first 20 years of its existence, SCAR changed very little. The membership was stable during this period, most of the delegates and working group members were experienced veterans of countless IGY and post-IGY Antarctic scientific expeditions, and most of the matters addressed were scientific or logistic. Since the middle to late 1970s, however, SCAR has undergone considerable change in its membership, in its constitution, and in its agenda.

The number of nations with SCAR delegates has increased from 12 to 18, with more to come. The number of delegates representing other ICSU bodies has increased also, although many of these delegates are also delegates from national committees. In addition, the cadre of alternate delegates has swelled the number of individuals in attendance at SCAR meetings to more than twice what could be expected in the early days. Also, women are now active in Antarctic science, a development that was unheard of during the IGY.

Many of the IGY veterans who were active in leadership roles in SCAR during its first two decades either have retired or are about to retire. It is therefore inevitable that SCAR will have to replace the old generation of SCAR scientists with a new generation. This may be traumatic for some who resist change wherever it occurs, but the new generation is ready and waiting to assume important roles in the affairs of SCAR.

To accommodate nations that aspire to membership in SCAR but have not yet satisfied SCAR's requirements, SCAR created a new class of members at XIX SCAR in 1986. These 'associate members' consist of nations that are gearing up for ongoing Antarctic programmes but may be several years away from their implementation. Countries accorded the status of associate member are able to participate fully in SCAR, as non-voting members, so that their plans for Antarctic operations will be enhanced by what they might learn from their discussions with other SCAR members. In 1983, the meeting of the representatives of the Antartic Treaty drew up the following recommendation (XII–8):

'*Recognising* that SCAR comprises a unique assemblage of knowledge and expertise in Antarctic scientific fields;

'*Noting* with appreciation the advice provided to the Antarctic Treaty Consultative Parties by SCAR in response to various requests;

'*Being aware* under its Constitution SCAR is charged with "furthering the co-ordination of scientific activity in Antarctica, with a view to framing a scientific program of circumpolar scope and significance";

'*Recommended* to their Governments:

'That they consider in the light of its expertise and great assistance any request that may be made by their national committees for additional funding to meet costs to SCAR of responding to requests for advice by the Antarctic Treaty Consultative Parties.'

Perhaps the greatest challenge to SCAR lies in its ability to deal with other international bodies and to address non-scientific issues without compromising the distinction between science and politics. SCAR has adhered rigidly to this distinction in the past and has been well served by so doing. But the line between science and politics has become more finely drawn, and SCAR must exercise constant vigilance to avoid becoming tangled in policy matters that, while they may relate to scientific activities, are the business of the consultative parties that administer the affairs of the Antarctic Treaty and related agreements.

SCAR must recognize that the scope of its activities will be broadened in future years. It has already responded through ICSU to a request from the United Nations for information in connection with the UN discussion on the question of the Antarctic. Along with this broadening of its agenda, SCAR will have to become more responsive to groups and organizations that have developed an interest in Antarctic affairs. SCAR can no longer function exclusively as a closed group whose members speak Antarctic jargon to one another at SCAR meetings and within the confines of the working groups.

A significant step in this direction was taken when SCAR joined with the International Union for the Conservation of Nature and Natural Resources in April 1985 in the sponsorship of a symposium on scientific requirements for Antarctic conservation. Further opportunities for joint sponsorship will undoubtedly be forthcoming from other organizations in the future, and SCAR must measure each such request against its basic mission. Moreover, SCAR must avoid being drawn into a position of advocacy, no matter how tempting some of these positions might appear to be.

As an organization, SCAR is an advocate only of the continuation of high-quality scientific research programmes in the Antarctic in accordance with the words in the preamble of its constitution. So long as SCAR can keep this mission in the forefront of its activities, and so long as its scientists can maintain strong programmes of high scientific merit, SCAR will continue to flourish as the only international body dedicated solely to the advancement of knowledge on this unique area of the planet Earth. There is nothing in SCAR's past or present behaviour to indicate that it will deviate from the mission that it established for itself at The Hague more than a quarter of a century ago.

James H. Zumberge
President of SCAR 1982–6
University of Southern California
Los Angeles
September 1986   *California, USA*

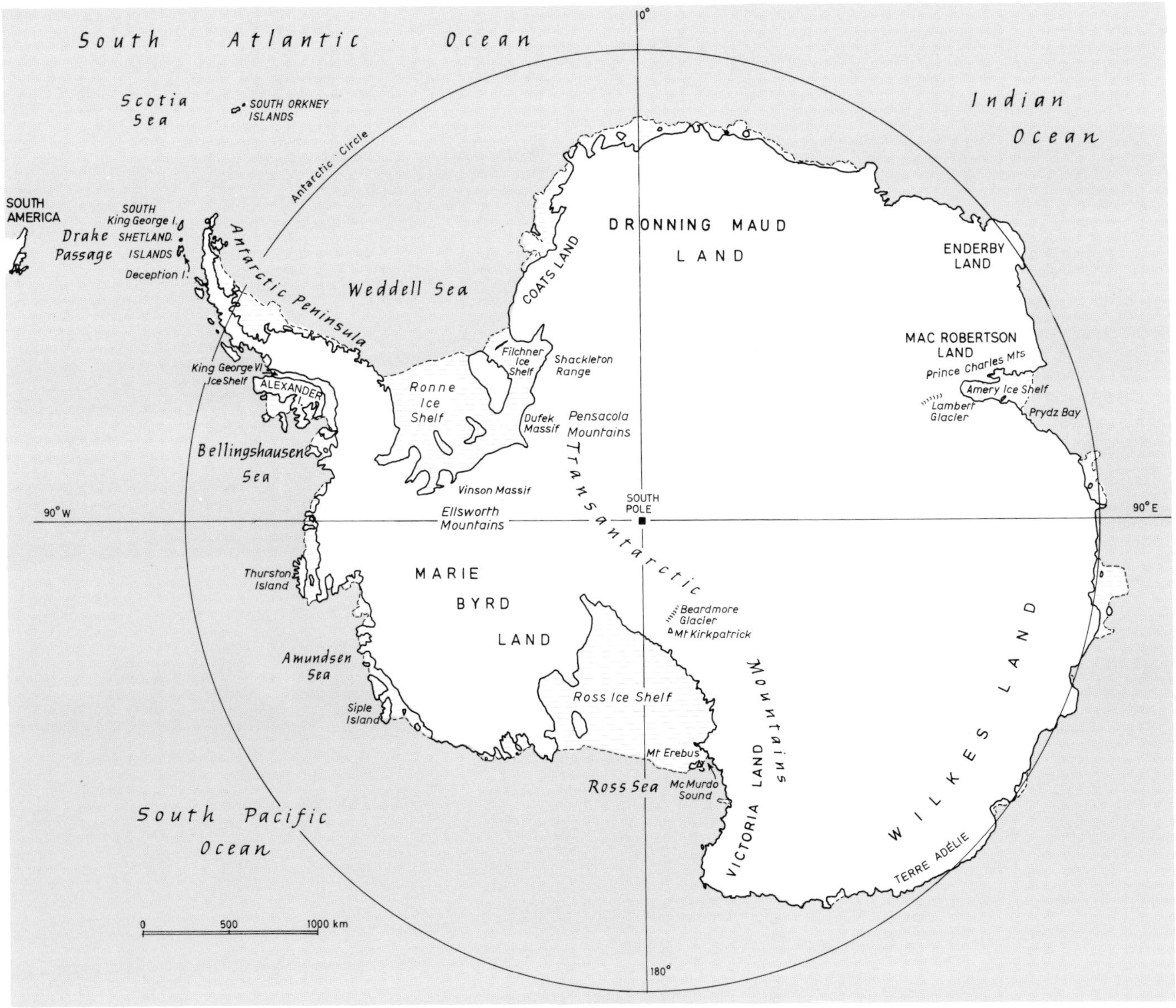

The Antarctic continent

Map labels

South Atlantic Ocean

Indian Ocean

Scotia Sea

SOUTH ORKNEY ISLANDS

Antarctic Circle

ENDERBY LAND

MAC ROBERTSON LAND

Prince Charles Mts

Amery Ice Shelf

Prydz Bay

Lambert Glacier

DRONNING MAUD LAND

COATS LAND

WILKES LAND

TERRE ADÉLIE

VICTORIA LAND

Shackleton Range

Filchner Ice Shelf

Dufek Massif

Pensacola Mountains

SOUTH POLE

Transantarctic Mountains

Beardmore Glacier

Mt Kirkpatrick

Mt Erebus

McMurdo Sound

Ross Ice Shelf

Ross Sea

Weddell Sea

Ronne Ice Shelf

Vinson Massif

Ellsworth Mountains

MARIE BYRD LAND

Antarctic Peninsula

King George VI Ice Shelf

ALEXANDER I

Bellingshausen Sea

Thurston Island

Siple Island

Amundsen Sea

SOUTH AMERICA

Drake Passage

SOUTH SHETLAND ISLANDS

King George I

Deception I.

South Pacific Ocean

90° E

0°

90° W

180°

0   500   1000 km

First impressions and the setting

SOME PERSONAL IMPRESSIONS

Cold earthless land, with immense ice islands which are continually separating in the summer, and are made by prevailing westerly winds almost to girdle the earth, and are evidently the cause of the very low temperatures.

James Weddell, *A Voyage towards the South Pole performed in the Years 1822–24* (1825), Longman Hurst Rees Orme Brown and Green, London.

A low white line, extending as far as the eye could see to the eastward. It presented an extraordinary appearance and proved at length to be a perpendicular cliff of ice . . .

James Clark Ross, *A Voyage of Discovery and Research in the Southern and Antarctic Regions, 1839–43* (1847), John Murray, London.

[Imagine] an immense city of ruined alabaster palaces . . . with long lanes . . . winding irregularly through them . . .

Charles Wilkes, *Narrative of the United States Exploring Expedition during the Years 1838, 1839, 1840, 1841, 1842* (1856), G. P. Putnam, New York.

Solid pack ice in the Weddell Sea

As we drew closer, the coast assumed a most formidable aspect. The most striking features were the stillness and deadness and impassibility of the new world. Nothing around but ice and rock and water. No token of vitality anywhere; nothing to be seen on the steep sides of the excoriated hills.

Louis Bernacchi, *To the South Polar Region* (1901); Hurst and Blackett, London.

What an evening! The sun is high in the heavens in spite of the late hour. Over all this mountainous land of ice, over the mighty Barrier running south, there lies a bright, white, shining light, so intense that it dazzles the eyes. But northward lies the night. Leaden grey upon the sea, it passes into deep blue as the eye is raised, and pales by degrees until it is swallowed up in the radiant gleam from the Barrier.

Roald Amundsen, *The South Pole* (1912), John Murray, London.

Let us follow the narrow sledge-tracks that the little black dots of dogs and men have drawn across the endless white surface down there in the South—like a railroad of exploration into the heart of the unknown. The wind in its everlasting flight sweeps over these tracks in the desert of snow. Soon all will be blotted out.

But the rails of science are laid; our knowledge is richer than before.

Fridtjof Nansen, in his preface for Roald Amundsen's *The South Pole* (1912), John Murray, London.

Beyond this flood a frozen continent Lies dark and wilde, beat with perpetual storms of whirlwind and dire hail, which in firm land Thaws not, but gathers heap, and ruin seems of ancient pile, all else deep snow and ice.

J. G. Hayes, *Antarctica: a Treatise on the Southern Continent* (1928), Richards Press, London.

... A continent emerged. In majesty of mighty mountain, peak and precipice, Antarctica was seen. The human race records but once discovery so great, so all embracing.

J. G. Hayes, ibid.

Camp in the Thiel Mountains

The thing we had come so far to see was before our eyes, a far flung reach of lifted ice, stretching east and west as far as the eye could see ...

Richard E. Byrd, *Little America: Aerial Exploration in the Antarctic, the Flight to the South Pole* (1930), G. P. Putnam, New York.

The scene, as we spiralled down, was one of wondrous beauty. An unbroken stillness, save for the hum of the propeller. The Barrier cliffs and slopes diffused the most exquisite colors, which changed and shifted as we watched. The lofty arch of sky was a clear blue, with friezes of perfectly stationary cloudlets, some rose, some mauve. A few icebergs glittered on a sea washed with gold; and in the west a range of the most beautiful mountains I have ever seen lifted purple peaks in tantalizing mirage.

Richard E. Byrd, ibid.

The view of an Antarctic ice shelf from seaward is hard to describe. Along the top it extends as level as the palm of one's hand for many miles. Its perpendicular edge is a sheer ninety degree drop into the sea ...

Paul W. Frazier *Antarctic Assault* (1958) Dodd, New York.

You gasp as a −29°F breath of air crashes in upon your lungs in one short gulp. The sun above is a dazzling fireball; the clouds and overcast have blown away a scant thirty minutes before. The snow is blinding in its glittering glare, even through the heavy dark sunglasses that must be worn to prevent snow blindness. In the distance— thirty, fifty, one hundred miles away—the staggering eleven-thousand- and twelve-thousand-foot needle-tipped peaks of the Royal Society Mountains loom over the horizon, looking so close you think you can reach them in an hour.

Allyn Baum, *Antarctica, The Worst Place in the World* (1966), Macmillan, New York.

Now the scenery which previously had been a vast open expanse of white upon white becomes ever changing. As the plane heads due South, the

The white desert of the South Pole

unbroken plain of sea ice to the left remains flat and featureless as far as the eye can see from 30 000 feet. But to the right are the glorious peaks of the Admiralty Range, the Prince Albert Mountains and the Royal Society Range .... The coastline of the Antarctic.

Allyn Baum, ibid.

Brown peaks, frosted by snow and tinged gold by the brilliant sun, slam 10 000 feet up from the sea of ice ...

Allyn Baum, ibid.

At McMurdo Base there is a sign that says 'You have arrived in Antarctica ... The end of the world'.

Allyn Baum, *Ibid.*

Antarctica is the largest, highest, coldest and cruellest desert in the world.

E. Honnywill, *The Challenge of Antarctica* (1984), Anthony Nelson, Oswestry.

Supplying a field party on the Rutford Ice Stream

ANTARCTICA: THE SETTING

POSITION AND AREA

Antarctica, an isolated, glaciated continent, is 990 km from the southern part of South America, some 2000 km from New Zealand, 2500 km from Australia, and 3800 km from South Africa. With an area of 13.9 million km², the continent represents one tenth of the Earth's land surface. It is larger than the United States and Mexico combined, almost twice the size of Australia, and around 15 times the size of Nigeria and Tanzania (924 000 and 940 000 km², respectively).

ELEVATION

Antarctica has a mean surface elevation of 2000 m, three times the average height of any other continents.

MOUNTAIN PEAKS

Some of the continent's mountain peaks soar to 5000 m—the highest being the Vinson Massif (5140 m) in the Ellsworth Mountains of West Antarctica. In places, the mountains project 3000 m above the ice sheet.

The ice sheet flows outwards in all directions to the ocean The rate of movement increases from less than one metre a year in the interior to typically more than 100 m a year near the coast; higher rates occur at outlet glaciers, where the rate may exceed 1000 m a year.

Glaciological and oceanographic studies at the front of the Riiser-Larsen Ice Shelf, Dronning Maud Land

ICE SHEET

Antarctica is the most glaciated of all continents and is blanketed in 25–30 million km^3 of ice distributed in the large, interconnected ice sheets in East Antarctica, West Antarctica, the Antarctic Peninsula and in the ice shelves flanking the Ross and Weddell Seas.

The ice sheet covers more than 98 per cent of the continent and is on average 2450 m thick, and as much as 4750 m in places. The weight of the ice cover has, in places, depressed the land by as much as 1000 m and a third of the land surface is below sea level: the Wilkes and Aurora sub-glacial basins, for example, extend in places to well over 1000 m below sea level. The Byrd sub-glacial basin is 2500 m below sea level.

TEMPERATURES

Temperatures in January (summer) can range from slightly above freezing along the coast to −30°C on the high interior, while in July (winter) the same regions average −20°C and below −65°C. In July 1983, the temperature dropped as low as −89.6°C at the Soviet Union's Vostok Station.

SOLAR RADIATION

Despite the months-long 'night' that results from Antarctica's polar position, the daily total solar radiative energy received during the summer at the South Pole is about the same as that received in Equatorial regions. This is due to the continent's elevation, thin atmosphere, and the longer day.

Marine platforms, at sea level, 40 m and 155 m above sea level, on King George Island, South Shetland Islands. Such platforms have been abraded by seas rich in floating ice-blocks

Charcot Station, sunk low in the ice, at an altitude of 2400 m and 317 km inland, provided a winter station for three people

Most of the Sun's heat, however, is reflected back into space by the high reflectivity, or albedo, of the ice and snow.

WEATHER

The weather around Antarctica is characterized by frequent depressions that travel predominantly between latitudes 50° and 70° South. More than six depressions can exist around the continent at any one time and can move eastwards at up to 85 km per hour (2000 km in a day). On the continent, cold air sinks from the high interior and flows downslope towards the coast, producing steady severe 'katabatic' winds. At times the katabatic flow is accelerated to strong blizzards, winds which can last several hours or even days and which may locally sometimes exceed 300 km per hour. Precipitation, mostly in the form of snow, occurs frequently over much of Antarctica, but it tends to be light. The interior plateau is a cold desert receiving less than 5 cm of water equivalent per annum. The air near the coast contains more water vapour and the annual snowfall there in some locations can be equivalent to over a metre of water.

ICE SHELVES

Towards the edge of the continent, the ice sheet generally becomes thinner and extends into the ocean, where it floats, forming vast ice shelves which add another 1.5 million km^2 to the area of the continent. The shelves can move seaward at speeds of up to 2500 m a year, and their average thickness is 475 m. The largest ice shelves are the Ross, Filchner, Ronne, and Amery.

SEA ICE

In winter, sea ice completely envelopes the continent and can extend as far north as 55° South. The area of ocean covered by ice can reach 20 million km^2—an area larger than the continent itself. Each year, in summer, 85 per cent of the pack ice disperses to the open sea and melts. It can move 65 km in a single day.

ICEBERGS

Antarctic icebergs that calve off the continental ice

Incoming snow storm

sheet and ice shelves are generally more numerous and much larger than their Arctic counterparts. A large number are many tens of km² in area, and a few can be up to 160 km long. They usually stand 30 m above sea level and they descend to 200–300 m below. When transported by ocean currents, they can move at a rate of between 10 and 20 km a day.

FRESHWATER CONTENT
Antarctica contains almost 90 per cent of the world's ice which constitutes about 75 per cent of the world's store of fresh water.

LIFE
Life forms in the Antarctic are concentrated primarily in coastal lakes, which are poor in species and low in productivity, or in the surrounding seas, which are relatively rich in species.

METEORITES
Between a third and a half of all the meteorites available for study in the world have been recovered from Antarctica. In four austral summers, nearly 4000 specimens of meteorites were found, including one identified as coming from the Moon and another possibly from Mars.

2

Why are scientists interested in the Antarctic?

MOTIVATION

Mankind has long been fascinated by the polar regions. The motives for acquiring knowledge of both the Antarctic and the Arctic have been many and include personal, scientific, economic, strategic and political ones. Such motives have led explorers, scientists and governments to be interested in Antarctica for the study of Earth, and the atmospheric, biological, physical, and oceanic sciences.

PURE ENVIRONMENT

The pristine quality of the Antarctic environment, as a result of its distance from sources of pollution as well as its special physical characteristics and geographic position, make the region uniquely suitable for a wide range of scientific research.

COMMUNICATIONS

The Antarctic is an ideal region for studying radio communications and the means by which it is influenced by solar-terrestrial phenomena. The magnetic field lines of the great magnetosphere surrounding Earth connect the northern geo-magnetic polar region with its Antarctic counterpart coming to Earth in the polar regions. Because of the shape of the magnetosphere, electrically charged particles and ionized gases (plasma) from the Sun are energized and directed into paths in the upper atmosphere above the poles. These particles, and those in the Earth's radiation belts, are responsible for magnetic storms which affect radiocommunication and also produce auroras. Such phenomena can be studied more easily from continental Antarctica than from the seas or floating ice of the Arctic.

AURORAS

If phenomena such as auroras could be better understood, they might help us to understand other phenomena in plasma physics, perhaps providing the clues on how scientists could harness fusion plasma processes to produce energy for humanity.

NATURAL RADIO SIGNALS

Natural electromagnetic disturbances generated in the Earth's magnetosphere by lightning in the

Checking out a remote automatic geophysical observatory

Aurora at Amundsen-Scott Station, South Pole. The geodesic dome that covers the American Station is seen at the right

Northern Hemisphere are carried southwards in natural waveguides along the Earth's magnetic field lines. The effects may be the amplification of these disturbances, the precipitation of charged particles, and the production of X-rays and various optical effects in the upper atmosphere. Such disturbances can cause, over a wide range of latitudes, excessive voltages in power transmission lines which may lead to power blackouts.

ANTARCTIC WEATHER AND CLIMATE

Antarctica exerts a profound influence on oceanic and atmospheric circulation in and beyond the Southern Hemisphere, and thus on the global weather and climate. The Antarctic ice sheet acts as a major heat sink in influencing climate and acting as an important control on world sea levels. If we are to understand the Earth's climate, we need to understand the atmospheric heat balance over Antarctica, the heat exchange between the Southern Ocean and the atmosphere, and the processes that can modify them.

ANTARCTICA AND SEA LEVEL

So much water is locked up in the huge Antarctic ice sheets that if they were to shrink or to melt, the level of the world seas would be drastically raised (perhaps by as much as 70 m) and many cities around the world drowned. The major part of the ice sheet is very stable because of its size and low temperature. However, concern is growing that ice sheets near the Antarctic Peninsula may suffer significant melting in the middle to latter parts of the next century.

CLIMATE CHANGE AND POLLUTION STUDIES

The ice sheets contain 'fossilized atmospheres' and are storehouses of information. Cores drilled from the ice sheets provide evidence of past climatic conditions and simultaneous changes in the composition of the atmosphere that may reflect the underlying causes of the climatic conditions. Ice cores reveal changes that have occurred over the past 10–200 000 years; changes in temperature,

A cliff of glacial stratigraphy

carbon dioxide and ozone levels, and atmospheric pressures. The ice sheets also store dust from distant and local volcanic eruptions, extra-terrestrial particles, pollen, pollutant residues, and fallout from atmospheric nuclear bomb tests.

SOUTHERN OCEAN

The Antarctic, or Southern, Ocean is the only great ocean whose East-West extent is not limited by continental land masses. To the south is Antarctica, to the north is the Antarctic Convergence, where cold Antarctic waters sink beneath the warmer waters of the Atlantic, Pacific, and Indian Oceans. Its area is twice that of the Antarctic continent. The circumpolar currents of West Wind Drift and East Wind Drift and the pack-ice, which covers half the Southern Ocean in winter, are characteristic features of the seas surrounding the Antarctic continent. Half the cold bottom waters of the world's oceans originate around Antarctica.

ANTARCTIC BIOTA AND KRILL

Antarctic waters support fewer species of animals

than tropical waters, but with larger populations. Biological productivity varies greatly in space and time, and can reach very high values. An important organism in the centre of the food chains in Antarctic waters is the shrimp-like krill, *Euphausia superba*. Krill has a wide distribution mainly in the zone which is covered by pack-ice in winter but which opens up in summer.

MARINE LIVING SYSTEMS AND STOCKS

A major scientific input to understanding the vital role of krill in the marine ecosystem came from the ten-year BIOMASS programme of scientific research that began in 1976. (BIOMASS stands for Biological Investigations of Marine Antarctic Systems and Stocks.) The principal objective of that programme was to gain a deeper understanding of the structure and dynamic functioning of the Antarctic marine ecosystem. The programme focused on the question 'how much krill is in the seas around Antarctica?' and 'how is krill

related to phytoplankton as its food source and to its predators such as fish, birds, and seals?'

ANTARCTIC GEOLOGY

Because some 98 per cent of the continent is covered by ice, many aspects of Antarctic geology are poorly known. For example, the structural and geological relationship between East and West Antarctica remains a major problem. There is very little information on the mineral resource potential of the continent.

PUZZLE OF A SUPERCONTINENT

In the geological past, Antarctica formed the nucleus of a great supercontinent, Gondwana, that also incorporated South America, Africa, Madagascar, India, and Australasia. Over the past 170 million years or so this supercontinent has been undergoing fragmentation and its pieces have been dispersed across the globe. A thorough investigation of Gondwana's history and the processes

Adélie penguins queue up to show their tags (above). The taking of blood samples is not the most dignified event for a penguin (left)

which led to its break-up are fundamental to our understanding of the forces which have shaped the surface of the planet, the distribution of animal and plant life, and the development of the southern polar environment.

BIOLOGICAL ADAPTATIONS

The harshness of the continent's cold and dry climate has many lessons to offer, including ways in which some Antarctic species have developed extraordinary evolutionary adaptations to the extreme environment. Antarctic fish have evolved a kind of natural antifreeze that prevents their blood from freezing, and some terrestrial organism must withstand conditions as harsh as those to be found in a deep freeze. Metabolism of all Antarctic marine organisms is adapted to temperatures around 0°C, approximately the freezing temperature of water.

HUMAN ADAPTABILITY

Because Antarctica is the most isolated continent on Earth and has no indigenous population, the scientific and technical groups who come from temperate and tropical regions form an ideal population in which to study human adaptability.

3

First studies of the Antarctic to the Treaty

INTRODUCTION

This is not the place to contemplate which nation or seafarer should be given the dubious honour of being recognized as the first to sight the Antarctic continent, let alone its ice or seas. Suffice it to say that among the early voyagers from the 16th through the 19th centuries were Dirck Gherritsz, Pedro Fernando de Quiros, Luis de Torres, Edmund Halley, Thaddeus Bellingshausen, James Cook, James Ross, Jean-Baptiste Bouvet, James Weddell, John Biscoe, Nathaniel Palmer, Charles Wilkes and Dumont d'Urville. Many of their names have been given to regions, places and features of the great southern continent.

Matthew F. Maury was one of the first to attempt to stimulate interest and international co-operation in the scientific exploration of the Antarctic. In 1860, Maury addressed a letter to the principal maritime nations asking them to participate in such a study, but without any positive results. The next attempt was made in 1871 at the First International Geographical Congress by Georg von Neumayer, who brought back the idea to successive Geographical Congresses until, in 1895, the Sixth Congress recorded its opinion that 'the exploration of the Antarctic Regions is the greatest piece of geographical exploration still to be undertaken ...' and recommended scientific societies to urge that scientific exploration be undertaken before the end of the century.

Henryk Bull and Carstens Borchgrevink were aboard the ship *Antarctic* which entered the Ross Sea and allowed a landing—the first undisputed one—on the Antarctic continent in January 1895. This was at Cape Adare. Thus began a new major period of scientific exploration of Antarctica. The expedition of Adrien de Gerlache on the *Belgica* in 1897 was the first to overwinter in the Antarctic ice, during the course of which the *Belgica* drifted from 80° to 102° West. The Belgian expedition was closely followed in 1898 by the *Southern Cross* led by Carstens Borchgrevink which left a party of 10, the first to overwinter on the Antarctic Continent.

INTERNATIONAL ANTARCTIC
CO-OPERATION

These two expeditions were followed, in the period 1901–4, by five scientific expeditions which

E. A. Wilson's sketch of aurora during the British Antarctic Expedition of 1901–4, as seen from winter quarters (site of present McMurdo Station)

Early physiological experiments with penguins

returned with a wide range of observations from the atmosphere down to the depths of the Southern Ocean: the *Antarctic, Discovery, Français, Gauss,* and *Scotia* led by Otto Nordenskjold, Robert F. Scott, Jean Charcot, Erich von Drygalski and William S. Bruce respectively. In 1903 the *Scotia* took a party of Argentine meteorologists to the South Orkney Islands. The party was relieved in 1904 by the *Uruguay* which landed a new group of meteorologists and rescued the Swedish expedition when their ship, the *Antarctic,* was lost in 1903. Other expeditions followed in the early 1900s, such as Ernest Shackleton in *Nimrod* in 1908 and *Endurance* in 1914/15, Robert Scott in 1910–12, Wilhelm Filchner in *Deutschland* in 1911/13, Douglas Mawson, who led the Australasian Antarctic Expedition of 1911/14, Jean Charcot in *Pourquoi Pas?* in 1908/10, the cruises of *Discovery* between 1925 and 1934, and the American expeditions under Admiral Byrd in the 1930s. The British–Australian–New Zealand expeditions of 1929/31 and the Norwegian–British–Swedish expedition of 1949 were the first multinational ventures of the modern age.

THE INTERNATIONAL COUNCIL OF SCIENTIFIC UNIONS

The International Council of Scientific Unions (ICSU) is a federation of scientists grouped together in 20 International Scientific Unions covering the whole field of natural sciences from astronomy to zoology. The specialist unions are interested in scientific research ranging from the outer limits of space to the liquid magma beneath the Earth's crust. The council was created in 1931 to encourage international scientific activity for the benefit of humankind. ICSU and the unions have member academies, research councils and scientific associations in more than 120 countries.

Over the past quarter century, ICSU, usually through specially established scientific committees or inter-union commissions, has organized a wide range of international co-operative programmes covering diverse fields of biology, Earth sciences, atmospheric sciences, solar-terrestrial physics, etc. The first of these was the International Geophysical Year (1957/58).

International Geophysical Year. The role of science in the Antarctic really came to the fore when a formal recommendation was submitted in 1950 to ICSU that a third International Polar Year be organized at the time of the next maximum of solar activity in 1957/58. Earlier Polar Years were in 1882/83 and 1932/33. The proposal included the following: 'It is also assumed that the Antarctic region would receive its full share of attention.' Because the natural phenomena to be studied were not confined to the polar regions but required study throughout the world, the programme became global and the name was changed in 1952 to the International Geophysical Year (IGY). ICSU established a Special Committee for the International Geophysical Year (CSAGI from the initials in French) to organize the programme. At the first meeting of the CSAGI, attention was drawn to the desirability of establishing stations on the mainland of Antarctica. This followed closely the pattern developed in the First and Second International Polar Years, where attention had been drawn to the importance of studies in high latitudes in the Southern Hemisphere, although during the International Polar Years no Antarctic stations had been established.

The first Antarctic conference organized by CSAGI was held in Paris in July 1955. The chairman of the conference, Georges Laclavère, emphasized that the overall aims of the conference were exclusively scientific, and that it was not concerned with financial and political questions. This statement was endorsed by the conference and has been the guiding spirit for the Scientific Committee on Antarctic Research, established under this name in 1958, in all of its subsequent programmes and activities.

ANNÉE
GÉOPHYSIQUE
INTERNATIONALE
1957
1958
INTERNATIONAL
GEOPHYSICAL
YEAR

Symbol of the International Geophysical Year, 1957–58

On the recommendation of the Fourth Antarctic Conference organized by CSAGI in 1957, the Executive Board of ICSU set up an *ad hoc* committee to examine the merits of further general scientific investigations in Antarctica. It resolved that there was need for further international organization of scientific activity in Antarctica, and recommended that ICSU should establish a committee to undertake this task. It also recommended that the committee should consist of a delegate from each nation actively engaged in Antarctic research, and representatives of the International Union of Geography, the International Union of Geodesy and Geophysics, the International Union of Biological Sciences, and the Union Radio Scientifique Internationale; each national delegate being authorized to bring with him to meetings advisers in various scientific disciplines, logistics, and communications. The continuation of scientific activity in Antarctic research was to be regarded as being inspired by the interest roused by the activities of the IGY, but in no way an extension of the IGY. In 1958, ICSU established the Special (later Scientific) Committee on Antarctic Research (SCAR).

SCAR. The first meeting of SCAR, in February 1958, prepared a plan for the scientific exploration of Antarctica in the years following the IGY. This plan has been amended from time to time. For the purpose of SCAR it was agreed that the 'Antarctic' shall be bounded by the Antarctic Convergence—the latitude between 40° and 60° South, where cold northward-moving surface water from Antarctica meets the warmer southward-moving

Stores and building materials landed at Coats Land for the advance party for the Royal Society's IGY Expedition, 1956

Royal Society Expedition
expands its station at Halley
Bay, 1957

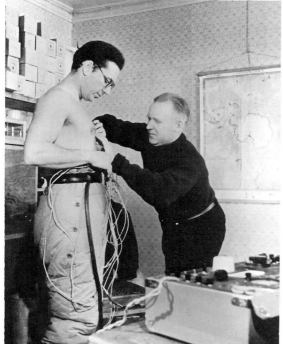

▲Testing the thermal properties of winter clothing at the
Soviet Antarctic station, Mirnyy, during the IGY

Japanese Antarctic Expedition for the IGY ▶

water from more temperate regions. Some subantarctic islands that lie outside the Antarctic Convergence, may be included in SCAR's area of interest: Ile Amsterdam, Iles Crozet, Gough Island, Archipel de Kerguelen, Macquarie Island, Prince Edward Islands, Ile St Paul and Tristan da Cunha.

For the first four meetings of SCAR, temporary scientific working groups were established. These were organized into permanent Working Groups with specific terms of reference after the fifth meeting, and at present there are eight: Biology; Geodesy and Cartography; Geology; Glaciology; Human Biology and Medicine; Logistics; Solid Earth Geophysics; and Upper Atmosphere Physics.

Specialist groups are also formed from time to time to address current research topics of special interest. At present, there are six Groups of Specialists on: Antarctic Climate Research; Antarctic Sea Ice; Southern Ocean Ecology; Seals; Evolution of Cenozoic Palaeoenvironments of the Southern High Latitudes; Structure and Evolution of the Antarctic Lithosphere.

The second meeting of SCAR, in August 1958, approved the publication of a bulletin containing a factual record of SCAR's activities, to be known as the *SCAR Bulletin* and to be published by the UK Scott Polar Research Institute and the Instituto Antártico Argentino. The meeting also encouraged further international co-operation, the exchange of scientific personnel and scientific documents, including maps. The subsequent exchange of scientists has been a notable, and valuable, feature of international co-operation.

In addition to the original institutional members of SCAR, the International Union of Geological Sciences, the International Union of Physiological Sciences, and the International Union of Pure and Applied Chemistry are now also members.

SCAR plays a part in all the international scientific projects involving worldwide observations. The committee is particularly concerned with the co-ordination of Antarctic contributions to these programmes, and with overcoming problems peculiar to the region. For example, SCAR will be much concerned with the Antarctic elements of the ICSU-World Meteorological Organization's World Climate Research Programme and has recently published *Antarctic Climate Research* (see Appendix 1, p 138).

THE ANTARCTIC TREATY

In December 1959 the Antarctic Treaty was signed at the conclusion of an international conference involving the 12 original signatory states. The first initiatives had come when the United States suggested, in 1948, a trusteeship arrangement for Antarctica under the United Nations Organization. The arguments against such a trusteeship led to a suggestion that the authority and control over the area should be vested in the hands of the directly interested parties. Negotiations on this suggestion continued with several governments until the United States distributed a note on 9 August 1948 to the governments of Argentina, Australia, Chile, France, New Zealand, Norway and the United Kingdom. That note suggested as a possible solution an idea for the promotion of scientific research in the area through agreement which would include some form of internationalization.

In September 1948, Julian Huxley, the first director general of the United Nations Educational, Scientific and Cultural Organization (UNESCO), suggested the creation of an International Antarctic Research Institute for the internationalization of scientific research in Antarctica under the aegis of UNESCO. The suggestion did not receive a favourable reception from the countries concerned, and it was not pursued.

In March 1949, the United States, in the preamble to a memorandum from the Department of State, emphasized the great value of scientific research in Antarctica and suggested that the parties seek an amicable and satisfactory solution which would avoid conflicts of sovereignty. Little progress was made with discussions until 9 June 1950 when the USSR addressed a note to the governments of Argentina, Australia, France, Norway, New Zealand, the United Kingdom and the United States proposing that international talks be opened forthwith, taking due account of the legitimate interests of all concerned. A series of factors, mainly political and territorial, led to a decreased interest in, and thus the shelving of, the proposal.

Moves by ICSU, to organize an International Geophysical Year revived interest in the idea for the creation of an international diplomatic regime for the region. In January 1956, the Prime Minister of New Zealand again suggested a United Nations trusteeship over the area and the establishment of Antarctica as a world territory. The suggestion was not adequately supported in the international fora, and it was not until September 1957, just after the start of the IGY, that the United Kingdom revived the earlier idea of the United States for the establishment of an international group of interested parties that would ensure the free development of science in Antarctica and that the area would not be used for military purposes. Although this initiative did not meet with immediate success, on 2 May 1958 the United States invited 11* other countries to seek an effective means of keeping Antarctica open to all nations, to conduct scientific or other peaceful activities there under joint administrative arrangements, and to participate in a treaty conference. On 4 June, the US Department of State announced that the 11 states had agreed to discuss the matter.

The Treaty Conference was held in Washington DC from 15 October until 1 December 1959 when the Antarctic Treaty was signed by representatives of the 12 nations participating. The main terms of the treaty include: Antarctica shall be used for peaceful purposes; freedom of scientific investigation in Antarctica and co-operation toward that end, as applied during the International Geophysical Year, shall continue. The contracting parties agreed that to the greatest extent feasible and practicable:

Information regarding plans for scientific programmes in Antarctica shall be exchanged to permit maximum economy and efficiency of operations;

Scientific personnel shall be exchanged in Antarctica between expeditions and stations;

Scientific observations and results from Antarctica shall be exchanged and made freely available.

In addition, it was agreed that in implementing Article III, on scientific investigation in Antarctica, every encouragement should be given to the establishment of co-operative working relations with those Specialized Agencies of the United Nations and other international organizations having a scientific or technical interest in Antarctica.

The fact that SCAR has continued to function effectively and smoothly and has established close and fruitful co-operation with other international organizations such as its sister Scientific Committees for Oceanic and for Space Research (SCOR and COSPAR) of ICSU, the World Meteorological Organization, the Food and Agriculture Organization, and the International Whaling Commission, is a clear testimonial to the validity of peaceful international scientific co-operation.

The Antarctic Treaty is open for accession at any time by any member of the United Nations Organization or by any other state which may be invited to accede to the Treaty with the consent of all the contracting parties. Parties that adhere are thus bound by the responsibilities of the Treaty but have full rights, equal to those of the original signatories, to share in the benefits. Voting rights at Treaty meetings are limited to states holding consultative party status acquired through substantial research activity in the Antarctic. A number of countries* have acceded to the treaty since 1959, the majority without joining the Consultative Group. The Treaty required that the contracting parties promote international co-operation in scientific investigation in Antarctica and that scientific observations and results be exchanged and made freely available. This means that the results are available not only to the contracting parties but to the scientific communities anywhere in the world.

In any future negotiations on the Antarctic, the essential points to be maintained are that the area shall be used for peaceful purposes; its uniqueness as a natural ecosystem shall not be jeopardized, and freedom of scientific investigation and international co-operation shall be maintained. Unfortunately, some of the recently emerged interests in Antarctica and suggestions to apply the concepts of international commons and common benefit are based on expectations of riches to be shared. Riches there may be, but the costs of basic scientific research and of eventual exploitation would be high. What safeguards can be introduced to ensure utilization of the natural resources of Antarctica without negative effects on the natural ecosystem? How do we ensure that the deep concern for the Antarctic environment demonstrated by the contracting parties will be assured by some other treaty?

---

*Argentina, Australia, Belgium, Chile, France, Japan, New Zealand, Norway, South Africa, United Kingdom, and the USSR.

*At the Thirteenth Consultative Meeting 18 consultative parties attended and 14 non-consultative parties.

Although SCAR is not mentioned explicitly in the Antarctic Treaty, it has been referred to several times in recommendations of the regular Treaty Consultative Meetings, in particularly I–IV with regard to the free exchange of information and views among scientists participating in SCAR (see Chapter 15). Moreover, the activities of SCAR constitute the kind of activity contemplated in Article III of the Treaty, 'SCAR should be encouraged to continue this advisory work which has so effectively facilitated international co-operation in scientific investigation.'

SCAR has provided scientific advice to the consultative parties which have adopted a number of measures consequent on advice from SCAR. For example, at the Third Consultative Meeting in 1964, the Treaty consultative parties adopted a series of agreed measures on Antarctic conservation based on SCAR recommendations. In this same context, the Treaty nations have set aside certain parts of the Antarctic as 'specially protected areas' (SPAs) and other areas designated as 'sites of special scientific interest' (SSSIs) on the advice of SCAR. Most Antarctic Treaty Consultative Meetings request advice from SCAR on various diverse issues.

4

Logistics, transportation, and telecommunications

Prior to the IGY, there had been some exchange of information about certain aspects of logistics such as boat, sledge, tent and other equipment design, nutritional requirements, and the types of scientific observation required.

Co-operation in logistics really became essential in the planning phase for the IGY (see Chapter 3), when so many new stations and expeditions were being developed by some countries with little, if any, experience of the problems of working and surviving under Antarctic conditions.

Indeed, the IGY was the first time in history that the Antarctic was subjected to a concentrated, simultaneous effort by scientists and support personnel from so many countries (12). Many had little experience in polar operations, and few were knowledgeable about the extremes of climate likely to be encountered and hence were uncertain about the design and type of material and equipment best suited to their task. All countries assisted others wherever possible, especially with space on ships, and later with the construction of stations to house the research teams. Those few with knowledge advised others as best they could, but communication was difficult to maintain and expertise was not always available when it was most needed. Lives were lost in those early endeav-

ours, but important lessons were learnt which were passed on to others to avoid repetition of the fatal errors.

When, in 1958, SCAR countries agreed to continue international science in the Antarctic for a long period after the IGY, and when they signed the Antarctic Treaty in 1959, it became apparent that activities in the Antarctic should now be regarded as permanent and there was a need for more formality and co-ordination in logistic planning.

In 1960, in order to promote co-operation and to facilitate exchanges of information on all logistical matters, a Logistics Working Group was established within SCAR. The group exchanges information through regular meetings, and maintains continuing communication on a wide range of logistical matters through its secretary. Such information includes new developments in sea, air and surface transport, Antarctic buildings and services, telecommunications, field equipment and clothing. This information is not restricted to members of the group. It is, and has been, made available to any country wanting to be better informed on Antarctic matters, and is especially important to those countries considering mounting their own programmes but lacking knowledge and experience in polar operations.

Motor toboggans, or snow-scooters, provide efficient and rapid means of transportation over snow

Many members of the Logistics Working Group are directors or managers of offices administering national Antarctic programmes; hence they have major responsibilities in the implementation of their national Antarctic policy in compliance with the Antarctic Treaty.

In recent years, therefore, the group has given attention to many 'managerial' matters. These include the formulation of policies for the implementation of controls on tourism and private expeditions, the development of management plans for specially protected areas and sites of special scientific interest, and the implementation of measures to ensure that

US Coast Guard ice-breaker Polar Star opening a channel through sea ice in McMurdo Sound

Field planning and communications headquarters for a glaciological drilling camp

environmental impact from human activities in the Antarctic be kept to a minimum. The group has also considered the safety of operations likely to result from any possible future commercial development, such as exploration for, or exploitation of, mineral resources.

SCAR, in recognizing this need, has implemented an annual exchange of operational information between its members in advance of, and additional to, the formal Exchanges of Information through diplomatic channels under Article XII (5) of the Antarctic Treaty.

A journey over snow can incur major hazards: blizzards and winds at up to 200 km per hour, and rough terrain including crevasses

Despite the introduction of modern technology to operations in the Antarctic, the area still remains the most hostile on Earth to human activities. Fullest possible co-operation in all endeavours remains paramount for the safety of all personnel and for the success of their work. Timely and detailed exchanges of information between all active parties are vital to achieve this high level of co-operation. The Logistics Working Group of

The group has established subcommittees with appropriate terms of reference to give attention to matters of special and continuing concern. One was set up in 1974 for planning and co-ordinating a Co-operative Air Transport System in Antarctica. Its main task was the study of intercontinental access routes to Antarctica with possible feeder routes to extend air access to many other stations, while its long-term objective is to

Helicopters come into their own in unloading the materials necessary for an expedition as in this one to the Pointe Géologie Archipelago

establish a system of international air transport to and within Antarctica using available resources and facilities to benefit all SCAR operators.

A Telecommunications Subcommittee was established in 1978 to advise Antarctic operators on new technical developments, such as the use of satellites, and to standardize on equipment and operating procedures. This group produced the *Antarctic Telecommunications Guidance Manual* (SCARCOM) published in 1983 and is responsible for collating data for the annual updating of the *Operators' Handbook*.

In addition to regular meetings, the group holds symposia and workshops to provide opportunities to concentrate attention on logistical problems of special importance. For example, in Bremerhaven in 1984, in response to a request from the Antarctic Treaty, the group arranged a workshop on 'Antarctic Telecommunications' for the purpose of identifying methods of improving telecommunications systems and the flow of data of importance to the global community.

Without doubt, the work of the group has contributed significantly to the success of international science in the Antarctic. International co-operation in logistics and the free flow of information on new materials, equipment and methods of operating safely and efficiently in the hazardous Antarctic environment, now so well established, will be maintained and expanded wherever possible by the Logistics Working Group of SCAR to the benefit of all countries.

5

Geodesy and cartography in the Antarctic

Mapping in the Antarctic before the IGY had two distinctive phases: the period prior to 1945 and the period from 1945 to 1952. During the first period, some coastal areas and waters and the subantarctic islands were surveyed. A few air photographs were taken as early as the 1920s. The results of these operations tended to be published in reports of expeditions or sometimes in scientific journals. From 1945 until the IGY, the nations with interests in the Antarctic began to establish mapping programmes and to produce maps and charts on various scales.

Although cartography was not included in the official programme of studies during the IGY period (1957/58 with the build-up starting as early as 1955), the 12 nations engaged in Antarctic activities (Argentina, Australia, Belgium, Chile, France, Japan, New Zealand, Norway, South Africa, United Kingdom, the United States, and the USSR) undertook a considerable amount of surveying. Maps were needed for planning scien-tific research and for the navigation of aircraft and field parties operating away from their station.

A photographic technique, developed during the Second World War, but which is now obsolete, used three cameras, one vertical and two oblique. The cameras were triggered simultaneously, and the technique made it possible to cover, in a single shot, a vast area of the ground. Trimetrogon air photography, as it was termed, became extensively used for mapping coastal and mountain terrain, and by the end of 1959 perhaps 5 per cent of Antarctica had been covered by it.

Thus cartographers had at their disposal a large amount of material from which they could compile maps. But this required knowing accu-rately position, latitude and longitude, and the elevation above sea-level of a certain number of features identifiable on the air photographs. The widely-used method of triangulation was not prac-tical over the ice sheet because any survey marks would move slowly and continuously with the

Surveying on Anvers Island in the Palmer Archipelago

ice sheet. The only possible way to obtain the horizontal control was by astronomical determinations ('astrofixes') of latitude and longitude. Unfortunately, serious errors may be introduced by the irregularities of the Earth's figure to which the astronomical observations refer. This figure, the 'geoid', is the shape of the Earth as if the surface were everywhere at sea level. As such, the geoid is a surface of equal gravitational potential. The visible mass anomalies (mountains) of the Earth and invisible ones (density changes beneath the surface) cause irregularities in the geoid and can introduce very large errors in the distance calculated between control points. Therefore, a network of astrofixes can be used only for small-scale mapping at 1:500 000 or less. Determination of altitude was also a major problem for cartographers. Precise levelling over great distances was very difficult.

Another complexity is that the Antarctic continent has two surfaces that need to be surveyed: the surface of the ice that covers the continent, and the surface of the underlying bedrock. The knowledge of this second surface aids the understanding of the structure of the continent itself. In the early days, the approximate configuration of the rock contours below the ice could be obtained only by seismic soundings made along ground traverses. The advent of a wide range of new technology changed all this.

AGE OF THE SATELLITE

In the late 1950s there was no significant geodetic network in Antarctica, and it did not appear possible to create one by traditional methods. Since then, both active and passive sensors on spacecraft and the development of laser technologies have facilitated the solution of the problems of Antarctic geodesy and cartography.

Geodetic measurements can be made from photographs of a satellite, taken simultaneously at a number of rotations. Such 'optical' methods were extensively used in the 1960s. However, since about 1970, they have been superseded by electromagnetic methods. Although the new technology is more expensive and complex, and requires satellites that are emitting a signal (and are thus said to be 'active'), it makes less demands on personnel, and works just as well in the bad or cloudy weather conditions of Antarctica.

The TRANSIT System operated by the US Defence Mapping Agency Hydrographic and Topographic Center, is composed of five active satellites. The orbit of the satellites is polar, quasi-circular at an altitude of the order of 1000 km. The satellites emit signals of a very stable frequency which are received by the ground station. By virtue of the 'Doppler' effect, the frequency of the signal received differs from that emitted by a quantity proportional to the radial velocity of the satellite with respect to the station. During a single

Surveying in mountainous terrain near Scott Station

passage of a satellite, many such measurements can be made of the Doppler integral, and the station is defined by the intersection of the corresponding orbital tracks.

In the past decade, considerable improvements have been made to the receivers for the TRANSIT system. The weight and size of the receivers have been reduced, so they are portable and can be installed in minutes. The observation of a single passage of the satellite, which takes 10–14 minutes, can give a position to better than 30 m. If the observations are repeated and if certain precautions are taken in the computation, the accuracy can be improved to better than one metre.

Another advantage of the system is that it gives, at the same time and with the same accuracy, the distances of the station to the centre of the Earth (actually the centre of mass of the Earth). Then if the elevation of one station of the network above the geoid (mean sea-level) is known, and if a map of the geoid in the area is available, it is possible to compute the altitude of the station by comparing the results of the Doppler observations with the map.

The TRANSIT system has been extensively used in Antarctica for mapping and for positioning the sites of certain geophysical operations (gravity measurements, seismic soundings, core drilling, determinations of ice velocity, etc.) along traverses. It will be superseded by about 1989 by the Global Positioning System (GPS) developed by the US Navy, Army, and Air Force and other US Administrations.

The GPS system will consist of 18 satellites in six different orbital planes at altitudes of about 20 200 km. An observer anywhere on Earth, even in the polar regions, will always have four satellites above his horizon. The system will give the distances between the observer and four satellites at a given time. As the exact position of the satellite at these moments is known, it is possible to compute the position of the observer, his latitude and longitude, and the distance to the centre of the Earth.

More accurate Doppler positioning, which requires the measurement of many satellite passes, has also been used to determine the surface velocity of the ice. An example is the Australian exercise to determine the ice flow across a section of the 2000 m contour using Doppler stations at 50 km intervals along the section.

PROBLEMS WITH THE ICE SHEET*

The estimation of the relief of the Antarctic ice sheet and its volume has been a major problem since the beginning of Antarctic exploration. The introduction of radar-altimetry virtually solved the problem. Radar pulses emitted from a spacecraft and the minimum times of propagation of the signal reflected from the ground surface are measured with great accuracy. From these measurements, the velocity of the signal being known, the vertical distance between the pulse transmitter and the ice surface can be computed. The altitude of the surface is the altitude of the satellite less the altitude measured by radar. The distance of a satellite from the centre of the Earth, and hence its altitude, may be computed from orbital data.

The accuracy of the method depends on the roughness of the surface covered by the wave front when it meets the ground. If it is even and horizontal, such as calm water, the accuracy may be as good as 10 cm; but if it is rugged or steeply sloping, the accuracy is much less. The accuracy also depends on the type of equipment employed: if the beam emitted is very narrow, the accuracy is better.

Similar pulsed radar systems at lower frequencies than those used on satellites were designed and built by the Scott Polar Research Institute (UK), and later by the Technical University of Denmark in collaboration with the US National Science Foundation, to measure the thickness of the ice sheet from an aircraft flying over it.

The US-Danish-British system was put into operation in Antarctica in the late 1970s under the auspices of the US National Science Foundation. Up to 1980, half the Antarctic ice sheet had been surveyed. Some of the rest has been surveyed by independent national expeditions (Australia, Belgium, Germany (FRG), Japan, South Africa, the United Kingdom, and the USSR).

Maps showing the surface relief of the ice sheet, the sub-ice relief, and both have been published by several agencies.

A RANGE OF MAP PROJECTS

All SCAR nations have produced maps of the areas where they are active. Topographic maps have been produced at a wide range of scales ranging from large-scale maps in the vicinity of

* See also Chapter 7.

A US operated ski-equipped LC-130 used for radio-echo sounding surveys (note antennae beneath the starboard wing)

the research bases to the small-scale maps (1:10 000 000) covering most of Antarctica. But the largest number of maps are at the 1:200 000 or 1:250 000 scales. Some coastal areas are mapped at 1:1 000 000 scale in addition to their mapping at medium scales.

LANDSAT images are now available over 60 per cent of the area of the continent, cloud-free or with less than 10 per cent cloud cover. Since the first launch of LANDSAT in July 1972, several satellites of this type have been launched equipped with sensors which permit the precise identification of a number of ice features. On the early models the picture resolution was 80 m; on the latest model it had been improved to 30 m. Using LANDSAT imagery, photo-mosaics have been

A section through the ice sheet from the Ronne Entrance, Bellingshausen Sea, West Antarctica, to Colvocoresses Bay, East Antarctica. (From *Antarctica: Glaciological and Geophysical Folio*)

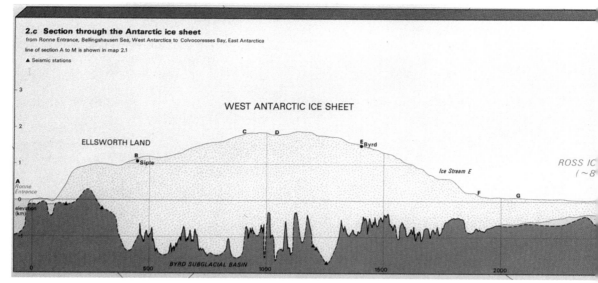

produced which cover some coastal portions of the continent.

WORKING GROUP ON GEODESY AND CARTOGRAPHY

From earliest days, it was recognized that accurate topographic maps were required by most sciences within the SCAR framework. Thus SCAR always emphasized the need for adequate map coverage of Antarctica as a whole, and particularly of all areas of ice-free rock and other identifiable detail. In the fulfilment of this requirement, SCAR has asked for full co-operation between member nations, and it approved, at its second meeting in 1958, the formation of a Working Group on Cartography which came into being in 1959. The name was changed in 1961 to that of Working Group on Geodesy and Cartography. As a result of the group's activities, SCAR made a series of formal recommendations pertaining to the subject. Some recommendations aim at ensuring co-operation between the SCAR member nations; others at standardization of mapping.

One of the first recommendations was a request for SCAR members to establish National Antarctic Mapping Centres and to urge its members to arrange automatic distribution to these centres of all maps of areas within the zone of interest of SCAR and also of data useful in the compilation of maps. It was also agreed that there should

be an exchange of information on unpublished mapping material and that agencies desiring copies could make special application to the originating source.

Another series of recommendations was aimed at ensuring a high degree of standardization in scales, symbols and projection used in the production of maps and atlases, and other recommendations dealt with aspects of geodesy.

In 1960, the Working Group on Geodesy and Cartography produced a catalogue of topographic maps, aeronautical charts, and hydrographic charts of the Antarctic published by member nations. The catalogue is kept up to date by periodical lists of new and revised maps. A revised version of the catalogue was produced in 1976 and a further edition is in preparation.

Data exchanged in accordance with the SCAR recommendations have enabled a number of general or thematic atlases of the Antarctic to be produced by agencies of SCAR members. In the 1960s and 1970s, the United States produced an extensive map folio series on Antarctica. Another spectacular production was the Soviet *Atlas Antarktiki* in two volumes. Volume I, first published in 1966, contained the cartographic and diagrammatic material. Volume II, first published in 1969, contained the text. It was a successful attempt to present in cartographic form the scientific knowledge of the Antarctic at the time of publication. The Arctic and Antarctic Research Institute of Leningrad planned and co-ordinated

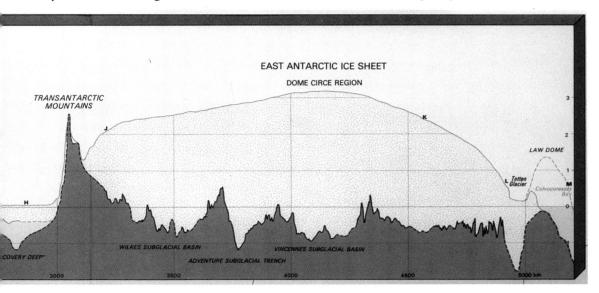

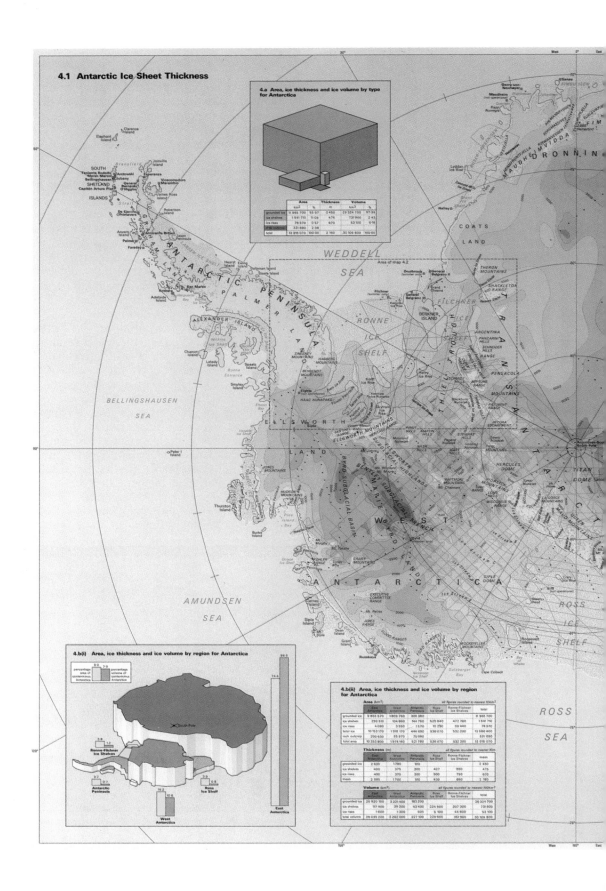

4.1 Antarctic Ice Sheet Thickness

4.a Area, ice thickness and ice volume by type for Antarctica

	Area		Thickness	Volume	
	km²	%	m	km³	%
grounded ice	11 965 700	85.97	2 450	29 324 700	97.39
ice shelves	1 541 710	11.09	475	731 900	2.43
ice rises	78 970	0.57	670	52 100	0.18
rock outcrop	331 890	2.38			
total	13 918 070	100.00	2 160	30 108 600	100.00

4.b(i) Area, ice thickness and ice volume by region for Antarctica

percentage area of continental Antarctica / percentage volume of continental Antarctica

Ronne-Filchner Ice Shelves — Antarctic Peninsula — West Antarctica — Ross Ice Shelf — East Antarctica

4.b(ii) Area, ice thickness and ice volume by region for Antarctica

Area (km²) — all figures rounded to nearest 10km²

	East Antarctica	West Antarctica	Antarctic Peninsula	Ross Ice Shelf	Ronne-Filchner Ice Shelves	total
grounded ice	9 855 570	1 809 760	300 380			11 965 700
ice shelves	293 610	104 860	144 760	525 840	472 760	1 541 710
ice rises	4 090		1 570	10 730	59 440	78 970
total ice	10 153 170	1 918 170	446 690	536 070	532 200	13 586 400
rock outcrop	200 630	55 970	75 090			331 890
total area	10 353 800	1 974 160	521 780	536 070	532 200	13 918 070

Thickness (m) — all figures rounded to nearest 10m

	East Antarctica	West Antarctica	Antarctic Peninsula	Ross Ice Shelf	Ronne-Filchner Ice Shelves	mean
grounded ice	2 630	1 780	910			2 450
ice shelves	400	375	300	427	650	475
ice rises	400		375	300	790	670
mean	2 585	1 700	550	430	660	2 160

Volume (km³) — all figures rounded to nearest 100km³

	East Antarctica	West Antarctica	Antarctic Peninsula	Ross Ice Shelf	Ronne-Filchner Ice Shelves	total
grounded ice	25 920 100	3 221 400	183 200			29 324 700
ice shelves	157 600	39 320	43 400	224 900	207 300	731 800
ice rises	1 600		1 300	5 900	44 600	53 100
total volume	26 039 200	3 262 000	227 100	229 600	251 900	30 109 800

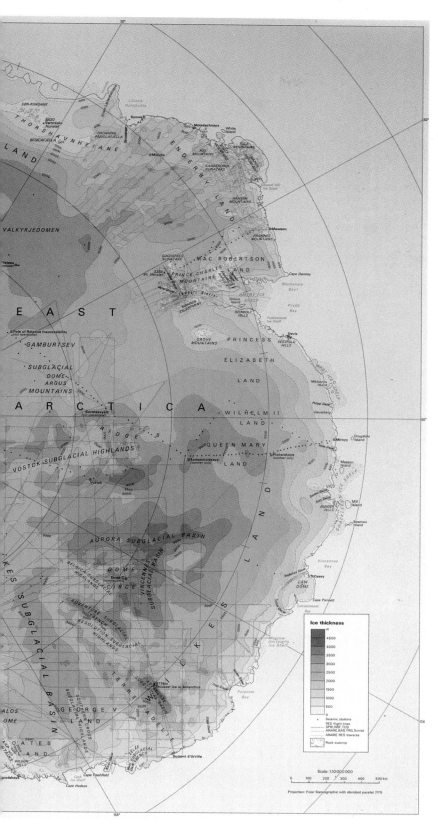

Sheet from *Antarctica: Glaciological and Geophysical Folio* showing the thickness of the continental ice

this atlas which was produced by Glavnoe Uprav-lenie Geodezii i Kartografii, Moscow.

In 1983, the Scott Polar Research Institute (SPRI) published the first part of *Antarctica: Glaciological and Geophysical Folio*. The purpose of this is to depict in a large-scale atlas format, the dimensions of the ice sheet and of the underlying bed rock directly measured or interpreted, principally from airborne radio-echo sounding and magnetic survey campaigns conducted by SPRI and others. Part of the folio is still in the course of compilation. In 1978, the Geographical Survey Institute of Japan published an atlas of eight magnetic maps of Antarctica. In 1982, Columbia Uni-

versity Press published the *Southern Ocean Atlas.*

Lists of Antarctic place names established by Argentina, Australia, Belgium, Chile, France, Japan, New Zealand, the United Kingdom, the United States, and the USSR have been used by the US Board on Geographical Names to compile a gazetteer of *Geographical Names of the Antarctic*, including approximately 12 000 names approved by this board, and 3000 unapproved variant names. It was published in 1981. Entries

Nimbus 5 satellite image of Antarctica during winter of 1974. The extent of the sea ice is clearly shown (false-colour, reds) and can be compared with the diminished sea ice in the previous summer (inset)

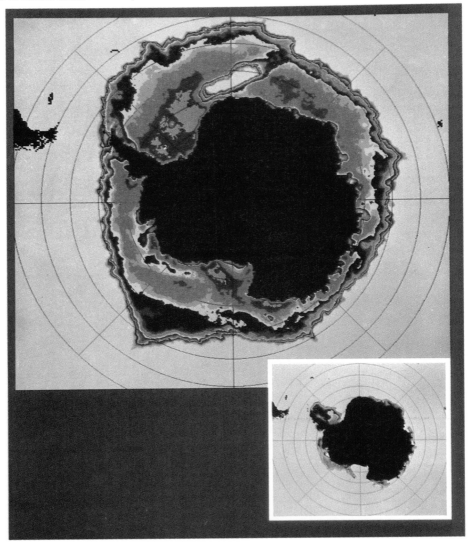

for approved names include location and description of the feature, and brief details on discovery and naming.

THE FUTURE

The foreseeable development of cartography in Antarctica during the next decade will be the extension of the map coverage of Antarctica at medium scale (1:250 000) and the replacement of conventional maps by hybrid forms in which symbols are superimposed on a satellite image background.

It can be expected that maps will improve in terms of quality, accuracy and completeness. The future use of the Global Positioning System will increase the density of the geodetic network and allow more accurate determination of latitude and longitude. Important improvements can also be expected in the determination of altitudes and in the drawing of contours of the ice cap, as new gravity and other measurements help to improve knowledge of the geoid and thereby improve the determination of altitudes obtained by the GPS at geodetic stations.

Progress can also be expected in radio-altimetric determination of the ice sheet. A group has been formed comprising scientists from a number of countries to undertake remote-sensing studies, in co-operation with the European Space Agency, with the specific aim of developing techniques for future satellite altimeter missions over ice-covered terrain.

It is likely that the resolution on LANDSAT imagery (30 m) will be improved to 10 m, permitting the identification of much smaller features. The French satellite SPOT, launched in early 1986, offers this quality; however, its images do not cover Antarctica south of 81° 40′ thus leaving uncovered an area of a radius of some 900 km around the pole. However, the arrangement of the two high-resolution imaging instruments aboard the spacecraft should enable photogrammetric treatment of the images and thereby the accurate mapping of Antarctica with a minimum of ground control.

The German Institut für Angewandte Geodäsie has developed new methods of digital image processing and great progress can be expected in image quality.

Interpretation of satellite imagery is of interest to a number of disciplines and is of paramount importance. Much research is going on in this direction in various institutions. To take two examples. First, areas of bare ice, not covered with snow, occur in many parts of Antarctica. They appear turquoise-blue and contrast sharply with the nearby 'white' snow cover. Meteorites have been found in certain 'blue-ice areas'. It is, therefore, important to delineate such areas on the maps by photo-interpretation.

Secondly, microwave images of polar regions are also a promising source of information. They permit the sensing of sea-ice, the determination of certain near-surface ice sheet parameters, and the inspection of the ocean surface through clouds and during the polar night. They provide a priceless tool for glaciological, oceanographic, and meteorological observations. The contribution of multifrequency microwave imagery to the protection of the Antarctic environment in the case of exploration and exploitation of mineral resources should not be underestimated.

6

Geology and Solid-Earth geophysics

SCAR has two subsidiary bodies concerned with the Earth sciences: the Working Groups on Geology, and on Solid-Earth Geophysics. Both working groups co-ordinate distinctive areas of activity, which produce research results that contribute to the broad geoscientific understanding of Antarctica, its continental margins, and the surrounding ocean basins.

Current geoscientific research in the Antarctic is first and foremost aimed at specific problems in the Earth sciences. These problems include determining and defining:

Antarctica's geological composition, structure, and evolution;

The geological relationship between East and West Antarctica;

The geological and palaeontological relationship to other continents and to the surrounding ocean basins;

The history of the development of Antarctica's current glaciation and the consequences of the glaciation;

Ice-encircled rocky peaks ('nunataks') of Beacon Rocks at Vestfjella, Dronning Maud Land

Earth processes in a polar, glaciated continental environment;

The relevance of Antarctica to global Earth science;

The mineral and hydrocarbon resource potential of the Antarctic region.

Outcrops of rock form only a tiny proportion (about 2 per cent) of Antarctica's land surface, and their occurrence is governed by the accidental process of geological preservation. Geological research in different areas of Antarctica involves different concepts, emphases, and techniques. For example, the geology of the Antarctic Peninsula (where the rocks are geologically quite young) does not have much in common with that of Enderby Land (where very ancient rocks are exposed). Antarctic geology is less concerned with continent-wide systems than are glaciology,

meteorology, and biology; on shore, at least, it is very much involved with deciphering the geological mosaic of the continent.

Geophysics is more of a continent-wide science. Its methods are used to study the physical properties of the Earth and to infer its structure and development. As such, geophysics has an important role in helping to improve our knowledge of Antarctic geology in general, and in resolving specific geological problems. Geophysical techniques make it possible to characterize some aspects of the subsurface geology without resorting to expensive direct sampling methods, that is, drilling.

PAST 30 YEARS IN ANTARCTIC EARTH SCIENCES

In the mid-1950s, knowledge of the geological composition, structure and evolution of

Field geophysics: inspecting magnetic measuring equipment on a Twin Otter aircraft (above). Later, the aircraft (below) returns with data on the contrasting magnetic properties of the rocks it has surveyed

Examining geological
sections in Vestfjella,
Dronning Maud Land

Antarctica was rudimentary, and broad-ranging reconnaissance programmes, some of which resulted in the discovery of whole mountain ranges, were predominant in the Earth sciences. Now, in the mid-1980s, few outcrops remain un-visited, and most geology programmes involve detailed or semi-detailed studies aimed at specific problems. Although much of Antarctic geology is now concerned with detailed research problems, reconnaissance is still the appropriate description for programmes relating to the parts of the continent covered in ice, and to the offshore continental margins. Direct geological knowledge of the continent's ice-covered areas is unlikely to improve strikingly for many years, although the geology may be inferred, in general terms, from geophysical data. Substantial advances are likely in knowledge of Antarctica's continental margins, which can be regarded as the main frontier areas of research in the Antarctic Earth sciences.

The general progress in Antarctic Earth sciences from reconnaissance to research has been accompanied by specific discoveries and advances including:

Recognition of Antarctica's central place in the configuration of land masses in the Southern Hemisphere through at least the Phanerozoic era of geological time—that is, the past 600 million years or so of the Earth's history;

Identification of geological similarities (and differences) between Antarctica and other land masses that were once adjacent to it, and recognition of Antarctica's significance for studies of the fossil flora and fauna of the Southern Hemisphere;

Confirmation of the fundamental geological differences between East and West Antarctica;

Elucidation of the geological composition, structure and evolution of the exposed parts of the metamorphic shield of East Antarctica;

Recognition of the Antarctic Peninsula as the focus of long-continued geological processes, resulting from relative movements of the Pacific Ocean plate beneath the margin of what is now Antarctica.

The description and study of unique geological features such as the Napier Complex in Enderby Land, the coal-bearing Beacon Supergroup rocks in the Transantarctic Mountains, the huge basic igneous intrusion (Dufek Intrusion) in the Pensacola Mountains, and the great extent and remarkably uniform composition of basic, dark-coloured igneous intrusive and volcanic rocks in the Transantarctic Mountains.

Recognition and description of the development, history, complexity, and variation of Antarctic glaciation in comparatively recent geological times, and of its profound influence on world climate, oceanography and sea levels.

The elucidation of glaciological and related sedimentological processes in Antarctica's continental polar glacial environment and the adjoining seas.

Definition of the shape of the land (topography) beneath the ice over wide areas of the Antarctic continent, and recognition of the topographic, geophysical, and glaciological differences between East and West Antarctica.

Description of Antarctic volcanoes and volcanic phenomena and recognition of their sig-nificance in geological and glaciological studies.

Recognition that the Antarctic ice sheet is a rich storehouse of meteorites, and the recovery and study of many such meteorites.

Very low seismicity of Antarctica.

A VAST SUPERCONTINENT

A number of geologists who worked with Antarctic expeditions in the early part of the 20th century recognized the geological similarity between Antarctica and Southern Africa, South America, Australia, and India. The discovery of the fossil seed fern, *Glossopteris*, in Antarctica was seen as good evidence in support of an earlier proposition that the various, now widely separated, land masses had once been linked together into a vast supercontinent called Gondwana. *Glossopteris* floras were already known from Africa, Australia, South America and India, and the original suggestion was that these land masses had been linked by land-bridges that had since disappeared. When it was suggested that the now separated continents had drifted apart, the concept of land-bridge links was no longer needed. The Gondwana supercontinent could be reassembled by retracing the trajectories of the various continents. When Gondwana was reassembled, the roughly circular shape of East Antarctica caused it to be termed the 'keystone' of Gondwana, with the other continents grouped

Pearse Valley, an ice-free valley in the McMurdo region of Southern Victoria Land

around it. From about 160 million years ago, this supercontinent progressively split apart and the various pieces drifted away from Antarctica to their present positions (Fig 6.1). Antarctica, the key piece, would seem to have remained almost stationary.

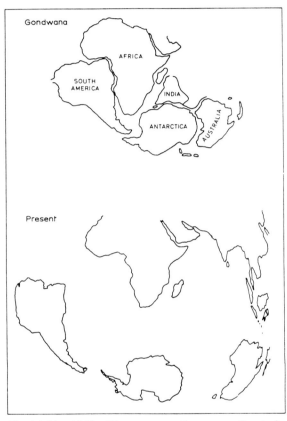

Fig. 6.1 About 160 million years ago, the supercontinent of Gondwana began to break up and the component parts drifted away from Antarctica

PLATE TECTONICS AND ANTARCTICA

Plate tectonics is the modern comprehensive theoretical concept that explains such things as continental drift, subduction and volcanism. It has formed the basis for a much improved understanding of the structure and mechanism of formation of the present surface of the Earth. Plate tectonics provides an explanation for the breakup of the supercontinent, Gondwana, and the formation of the present Antarctic continent.

Antarctica lies within a single lithospheric 'plate', with East Antarctica bounded only by

passive continental margins and presently active margins occurring only off the Scotia Arc to the north-east of the Antarctic Peninsula. Marine geomagnetic measurements have made it possible, by extrapolation, to assign an age to much of the ocean floor around Antarctica, and to reconstruct the broad development of the Antarctic margin and Southern Ocean during the break up of Gondwana. However, attempts to 'reconstruct' Gondwana as it was are thwarted by the problem of overlap between the Antarctic Peninsula and South America. Clearly, both land masses could not have been in the same place at the same time. To overcome this, geologists have speculated that West Antarctica is formed of a number of fragments (termed 'terranes') which have moved relative to each other. They may have rotated as well as moving laterally, which further complicates the story. Some plate tectonic models also suggest rifting between East and West Antarctica during early Tertiary times (about 60 million years ago), and possibly convergence along the line of the modern Transantarctic Mountains.

The plate tectonic history of the Antarctic region gives the timing of the opening of seaways between continental masses. These are related to the onset of major ocean current systems—for example, the opening of the Drake Passage between the Antarctic Peninsula and South America began about 29 million years ago, and that between North Victoria Land and Tasmania at about the same time. This allowed the eventual development of the Antarctic Circumpolar Current. Thus the plate tectonic history is relevant to the palaeogeography, palaeo-oceanography and palaeoclimate of the region.

AN APPARENTLY EARTHQUAKE-FREE AREA

The Antarctic continent is the largest apparently earthquake-free area on the Earth, and few earthquakes have been located routinely there by the Worldwide Standardized Seismograph Network (WSSN). More than 10 seismological stations have been operating in Antarctica since the IGY, and small local tremors recorded by them are commonly attributed to calving of the ice shelves or fracturing of the ice sheet. Small earthquakes, probably of volcanic origin, are associated with the active volcanoes at Mount Erebus, on Ross Island, and Deception Island near the Antarctic Peninsula.

The WSSN can detect almost all earthquakes

The crater of Mount Erebus, Ross Island, shows signs of periodic volcanicity

in the world with a magnitude greater than 5 on the Richter scale. Only three such earthquakes have been recorded from Antarctica, one in 1952, one of magnitude 4.9 in 1974, both originating in northern Victoria Land, and close to a major glacier and ice tongue. The third earthquake, in 1985, was reported to have occurred in Dronning Maud Land. Seismologists suggest that, although the 1974 event had characteristics resembling those of an earthquake generated by normal geological processes, it is likely to have been caused by movements within the ice. The 1985 event, in contrast, was attributed to normal tectonic processes and remains the only unquestionable Antarctic earth-tremor.

ANTARCTIC CRUSTAL STRUCTURE

The crustal structure of Antarctica was virtually unknown 30 years ago, prior to the IGY. The IGY studies focused on the dispersion of seismic surface waves from distant earthquakes detected at seismograph stations on Antarctica. These showed that the thickness of the Earth's crust beneath Antarctica was not uniform and that it ranged from thicknesses typical of the continental shield areas of northern America and Europe, to intermediate and thinner than normal thicknesses, perhaps more typical of continental rift regions such as the East African rift. As a result, it was proposed that Antarctica could be broadly divided into two parts, a large part with a thicker crust (East Antarctica) lying mainly in the Eastern Hemisphere and a smaller part (West Antarctica) with a thinner crust of uneven thickness. Gravity surveys during the late 1950s and early 1960s confirmed this broad division.

Since the surveys of the IGY, additional measurements, especially seismic refraction and reflection measurements, have defined the crustal structure in more detail in several locations. These deep seismic-sounding experiments have been carried out off the northern Antarctic Peninsula, in the south and east Weddell Sea, and in Dronning Maud Land, western Enderby Land, the

Two geophysicists check a seismometer on the flanks of Mount Erebus, while a helicopter waits to take them back to their station

Amery Ice Shelf region and McMurdo Sound. They confirm that thick crust (40 km) underlies East Antarctica and a thin (25 km) crust underlies the Ross Sea to Weddell Sea depression.

A DISTINCTIVE CONTINENT

The Antarctic continent is distinctive in being almost wholly covered by ice, essentially free from earthquake activity, and in apparently having had an almost fixed location for the past 200 million years or so (even though the fauna and flora were temperate in type during much of that time). Other distinctive geological features of the region include: the crustal fragmentation of West Antarctica, the Dufek Intrusion in the Pensacola Mountains, the local concentrations of meteorites on the ice sheet, the Napier metamorphic complex in Enderby Land, and such widespread geological entities as the Beacon Supergroup sedimentary rocks, and the Ferrar Group of igneous rocks, which intrude the Beacon rocks and overlie them.

The contribution of Earth science research in Antarctica to global geology must be measured in terms of such distinctive features. Antarctica's glaciation provides unique opportunities to study the effects, dynamics and processes of continental-scale glaciation like that which overwhelmed large areas of northern Europe and North America only a few tens of thousands of years ago. It also provides a guide to interpreting rocks that were produced in many places on Earth during glacial eras in past geological times, and insights into the climatic and oceanographic history of the Earth in geologically recent times. On the other hand, Antarctica's lack of earthquake activity remains enigmatic, being due perhaps to the continent's polar position, perhaps to its relative immobility compared with other continents, or as a consequence of glacial loading of the Earth's crust.

The rocks of West Antarctica record a long history of complex plate tectonic processes including subduction, ridge crest-trench collision and crustal fragmentation. It is possible that this region will provide a distinctive perspective to some fundamental concepts of geological processes in the Earth's crust.

The Dufek Intrusion poses rather than answers geological problems; but it is important to try to understand why and how such large masses of igneous rock are emplaced in the Earth's crust,

Walker Peak, in the Dufek Massif, shows distinctive layered igneous rocks

and to relate this understanding to Earth processes on a broader scale. The intrusion appears to be part of the larger-scale intrusion constituting the Ferrar, Tasmanian, and Karroo dolerites.

The discovery of concentrations of meteorites on the Antarctic ice cap has had a profound impact on meteorite and planetary geology studies. This is because the meteorites are very little weathered, and because the collections of the different kinds of meteorites may reflect cosmic abundances more closely than collections from other continents. While only four meteorites had been recorded in Antarctica up to 1969, over 6000 meteorite fragments have now been recovered, mostly from these areas.

Whereas the meteorites provide insights into extra-terrestrial processes and events, the Napier Complex rocks of Enderby Land have yielded information on processes that took place deep in the Earth's crust, under unusual conditions, early in the Earth's history. Studies of such rocks will yield information on the early development of the Earth's crust only 600 million years after the

estimated age of the origin of the Earth, and thus contribute, along with findings in other parts of the world, to a better understanding of the Earth's history and processes.

Research into the Beacon Supergroup and Ferrar Group rocks also contributes to an understanding of the Earth's history and processes although at a very different stage in the planet's evolution. For example, the Beacon Supergroup rocks have yielded rock types and fossils, including vertebrates, that provide evidence that Antarctica, Australia, India, South Africa, and other southern continents once shared very similar climatic conditions. The Ferrar Group rocks together with similar rocks in southern Africa, South America and Australia testify to the production and emplacement of huge volumes of molten rock of remarkably uniform composition over a very large area. This igneous activity took place at much the same time as the commencement of breakup of part of the Gondwana supercontinent. Study of the Ferrar Group rocks will clearly contribute to an understanding of the

One of the many large brown meteorites found in pristine condition in the Yamato Mountains, East Dronning Maud Land. This one weighed 5.5 kg and measured $20 \times 18 \times 10$ cm

Earth's geochemistry and magmatic processes in the crust and mantle.

THE CONTINENTAL SHELF
The edge of the continental shelf of Antarctica lies at an unusually great depth (500 m), and large areas are covered by ice shelves. Most of Antarctica's continental margin is a passive, or rifted, type with apparently only minimal sedimentation at the present time. Investigating the structure of the margin and the reason for the deep continental shelf will help to clarify our

Deploying a rig for photographing the sea-bed adjacent to an ice shelf

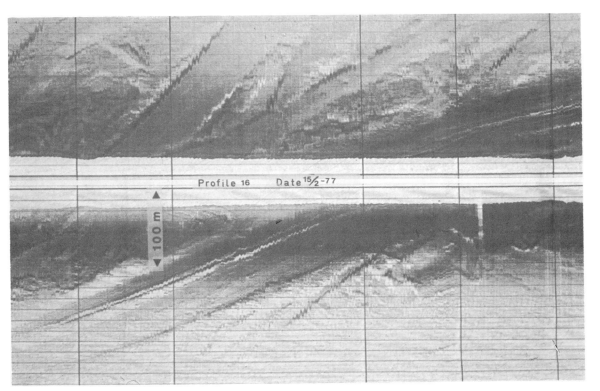

Profile 16   Date $^{15}/_2$-77

100 m

Side-scan sonar record of the
sea-bed revealing the 'plough
marks' left by a scouring
iceberg

Launching a large box sampler from ship deck
to collect seabed samples

A box corer at work over the side of a ship

Paying away a seismic cable
with floats from the stern of a
research vessel in the Weddell
Sea. The cable consists of
arrays of geophones to pick
up reflected sound waves
during seismic surveying. The
arrays can stretch up to 4 km

understanding of the formation of continental margins, shelves and marginal plateaux in Antarctica and elsewhere. In particular, data on the continent-oceanic crust transition and crustal stretching, thinning and rupture are of interest. Antarctica is a suitable place for these studies, as the relatively thin cover of sediments on its passive margin may make it easier using seismic stratigraphy to interpret the development of the margin with time.

SEISMIC SURVEYS

Multichannel seismic methods (both reflection and refraction) are well proven in marine and land areas for the study of subsurface geology. However, experience in this method in ice-covered areas of Antarctica is limited, especially where the ice is thicker than 200 m. Problems arise from high reflectivity at the ice-rock interface and in the

Deployment of a vibrocorer in the Weddell Sea to obtain cores of the sea-bed

high seismic velocity of ice, which also limits the effectiveness of seismic refraction methods. Nevertheless, it is considered that the seismic technique will be an important one for studying sub-ice geology and thus be particularly of value in Antarctica. Several parts of the interior of Antarctica are inferred to be underlain by sedimentary basins; good seismic data may enable Earth scientists to test these theories. High-quality seismic and aeromagnetic data would also be required for the detailed investigation of sites for drilling through the continental ice cap for both glaciological and geological research.

SATELLITES AND ANTARCTIC EARTH SCIENCES

Increasing use of remote (satellite) sensors for study of the Antarctic region is anticipated. Studies of long-wavelength geomagnetic anomalies have been undertaken using MAGSAT results, but higher spatial resolution data and higher latitude (polar) orbit observations are required. Satellite altimetry, which detects perturbations of the sea surface, can be used to investigate the deep structure of continental margins when sea ice prevents ship studies. It also provides considerable detail of the geoid. Gravity satellite missions will also contribute to the study of the shape of the Earth.

MINERAL RESOURCE POTENTIAL OF THE ANTARCTIC

As noted previously, research into the Earth sciences in Antarctica is primarily aimed at scientific problems, and this is the main reason why the scientists undertake their work often under difficult and sometimes dangerous conditions. However, one by-product of the research is information that must be taken into account when the Antarctic's potential for resources is considered. For a meaningful appraisal of the continent's petroleum and minerals potentials, it is vital that information resulting from Antarctic geological and geophysical research be assessed.

The development of the Antarctic continent, especially its relationship to the formerly juxtaposed continents, makes it possible to speculate about the likelihood of resources existing there. The location of zones of economic resources on the once adjacent continents can be used in attempting to predict where similar zones might occur in Antarctica. Furthermore, the history of breakup of Gondwana will provide information on the geological development of the continental margin during and after the breakup of Gondwana, and thus on the potential for mineral and petroleum resources.

Petroleum occurs in sedimentary rock sequences. On the Antarctic continent, most of the few sedimentary rocks that are exposed are unsuitable because of their composition, because they are too deformed, because they have been heated too much during deep burial, or because they are too thin. The continental shelves seem to hold more promise, not the least because they are more accessible, although some of the most interesting areas for possible sedimentary basins are covered by ice shelves. Thick sedimentary sequences have been identified in many areas along the Antarctic margins including the Ross Sea, Weddell Sea, and Prydz Bay continental shelves as a result of multichannel seismic studies, and a picture is emerging of at least seven major basins around Antarctica (see Fig. 6.2).

In the Weddell Sea and the Ross Sea, the sedimentary rocks may reach a thickness of more than 10 km. The former area, however, is characterized by some of the most notorious pack ice in the Antarctic. By contrast, the Ross Sea, north of the Ross Ice Shelf, clears itself of ice more or less every year. This information has been used to make a number of preliminary assessments of possible resources by analogy with other, better known, sedimentary basins and, more recently, by modelling the resource potential using available geophysical and geological data. These data are few, and much further work is required before any reasonable assessment of possible resources can be made. Even so, experience elsewhere has shown that assessments, such as these, made without any drilling information can be very speculative and uncertain. Furthermore, the sort of world economic situation that might lead to the exploration and exploitation of Antarctic petroleum seems a remote possibility. A number of formidable political, legal, technical and environmental obstacles would have to be overcome. Without doubt, the most important obstacle to development is, and will continue to be for a long time, economic. Certainly, recent (1986) fluctuations in the price of crude oil serve only to emphasize this view.

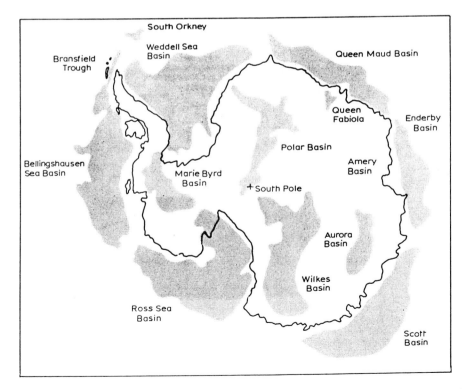

Fig. 6.2 Schematic diagram of the main sedimentary basins of the Antarctic

OTHER COMMODITIES

Coal of Permian age (about 250 million years old) occurs throughout much of the length of the Transantarctic Mountains. The coal is generally characterized by being thin, lenticular, and discontinuous, with a high content of ash and moisture. The grade is higher where the coal has been baked by later intrusion of dolerite. Much energy would be required to recover any Antarctic coal, which therefore must be viewed at present as an uneconomic resource, from the standpoint of both money and energy. Permian coal of higher (though still low) quality occurs in the Beaver Lake area of the Prince Charles Mountains.

Iron deposits occur in Antarctica, and especially in East Antarctica, where traces of banded iron formation are known from Enderby Land through to Wilkes Land, the largest occurrence being in the Prince Charles Mountains in Mac. Robertson Land. Here, at Mount Ruker, banded iron formations up to 70 m thick are interbedded with metamorphosed rock types. Geophysical surveys indicate that the iron formation extends for several tens of kilometres under the nearby ice sheet. There are also minor occurrences of iron-rich rock in Dronning Maud Land, in the Dufek Intrusion and in the Transantarctic Mountains. The iron content of these formations is, however, not high. If one bears in mind their remote location and the hazards and expense of exploration, mining and transport it appears that economic exploitation is extremely unlikely.

Copper, although only in minor mineralization, has been found during geological studies in the Antarctic Peninsula and links with mineralization in South America have been suggested. There are, however, important differences, including the observation that copper mineralization of the Andes does not continue to the southern Andes.

Within the very remote, inland Pensacola Mountains there are exposures of the Dufek Intrusion—a large, layered, basic igneous complex. These exposures are only a small part of the intrusion, however, and airborne magnetic and radio-echo sounding surveys indicate that it extends over an enormous area, maybe as much as 50 000 km².

Geologists have compared the structure and composition of the Dufek Intrusion (170 million years old) with that of the mineral-rich Bushveld Complex (2100 million years old) in South Africa. This has fuelled speculation on the resource potential of the Dufek Intrusion for deposits of metals of the platinum group and chromium. However, the difference in age between the two intrusions casts considerable doubt on such speculation, as do available geochemical data. More research is required before a reasonable assessment can be made. Present indications, however, are not encouraging.

Many types of mineral occurrence, important elsewhere in the world, cannot be expected in Antarctica. Residual deposits, such as bauxite, will have been destroyed by glaciation, as will various weathering enrichments, heavy mineral beach sands, and so on.

Manganese nodules and encrustations are known from widely scattered locations in the Pacific sector of the southern oceans. Generally, manganese nodules occur far from land, on the deep seafloor. Some reports suggest that the copper, cobalt, and nickel contents of the nodules from the Pacific Ocean are dependent on latitude. Those in equatorial regions contain more associated metals than those in higher latitude.

Ice, from many technical points of view, is of course, a mineral. There is no question that fresh water in the form of ice would be welcome in most arid countries. The estimated annual yield of icebergs in Antarctica is of the order of 1000 km³. The problem would be how to get the resource from where it is to where it could be put to use. The SCAR report *Possible Environmental Effects of Mineral Exploitation in Antarctica* states: 'The idea of using Antarctic icebergs has been carefully investigated and even tried in a small way. Theoretically, icebergs could be floated to any point accessible by water route with minimum depths of 200 m. . . . All the feasibility studies of the past few years have concluded that the operation of towing the icebergs was not beyond reason . . . [though] no unprotected iceberg would survive the journey from the Antarctic to low latitudes . . . yet to do so to southern Australia [would be] not beyond reason.'

It is doubtful that any metallic or non-metallic mineral resources in Antarctica will be worth exploiting for many years, unless world economic or political conditions change drastically.

7

'Land of mountainous ice'*

Ice on land and in the ocean is the single most distinctive feature of the Antarctic. The Antarctic ice sheet is in parts as much as 4.75 km thick and contains some 90 per cent of the world's ice and 70 per cent of the world's store of fresh water. Remote from any industrial centre, and most of it from both the ocean and exposed rock, the ice sheet forms the purest natural water on Earth. Beyond its hydrogen and oxgyen, the ice contains only a minute concentration of other elements— a few parts in 1000 million, which means that it is purer than laboratory analytical grade water. Much of Antarctica is technically a desert, and each year more than half its area receives in snowfall less than the equivalent of 10 cm of water. Essentially no melting of the snow occurs (except at the margins), even in the summer, and each year a new layer is added. As the snow layers sink into the ice sheet they become compressed and progressively thin under the weight of the overlying layers of ice, which causes the ice sheet to flow outwards to the coast.

As snow falls, it incorporates the fine aerosols from the atmosphere. These may include continental dusts, sea salts, volcanic particles, cosmic

* Thaddeus Bellingshausen.

Astrolabe Glacier, Terre Adélie, at the end of an Antarctic winter

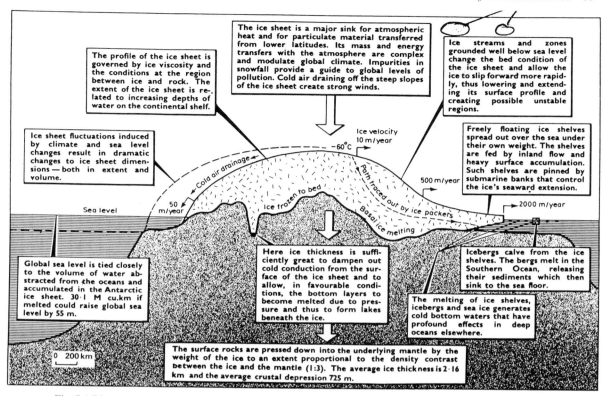

The profile of the ice sheet is governed by ice viscosity and the conditions at the region between ice and rock. The extent of the ice sheet is re-lated to increasing depths of water on the continental shelf.

The ice sheet is a major sink for atmospheric heat and for particulate material transferred from lower latitudes. Its mass and energy transfers with the atmosphere are complex and modulate global climate. Impurities in snowfall provide a guide to global levels of pollution. Cold air draining off the steep slopes of the ice sheet create strong winds.

Ice streams and zones grounded well below sea level change the bed condition of the ice sheet and allow the ice to slip forward more rapid-ly, thus lowering and extend-ing its surface profile and creating possible unstable regions.

Ice sheet fluctuations induced by climate and sea level changes result in dramatic changes to ice sheet dimen-sions — both in extent and volume.

Freely floating ice shelves spread out over the sea under their own weight. The shelves are fed by inland flow and heavy surface accumulation. Such shelves are pinned by submarine banks that control the ice's seaward extension.

Ice velocity 10 m/year

−60°C

Cold air drainage

Path traced out by ice packers

Basal ice melting

Sea level

50 m/year

Ice frozen to bed

500 m/year

2000 m/year

Global sea level is tied closely to the volume of water ab-stracted from the oceans and accumulated in the Antarctic ice sheet. 30·1 M cu.km if melted could raise global sea level by 55 m.

Here ice thickness is suffi-ciently great to dampen out cold conduction from the sur-face of the ice sheet and to allow, in favourable condi-tions, the bottom layers to become melted due to pres-sure and thus to form lakes beneath the ice.

Icebergs calve from the ice shelves. The bergs melt in the Southern Ocean, releasing their sediments which then sink to the sea floor.

The melting of ice shelves, icebergs and sea ice generates cold bottom waters that have profound effects in deep oceans elsewhere.

0 200 km

The surface rocks are pressed down into the underlying mantle by the weight of the ice to an extent proportional to the density contrast between the ice and the mantle (1:3). The average ice thickness is 2·16 km and the average crustal depression 725 m.

Fig. 7.1 Diagram of the Antarctic ice sheet, illustrating various interactions with other parts of the environment

dusts, and industrial pollutants. They become incorporated in the ice together with trapped bubbles of atmospheric air and are preserved indefinitely without further biological or chemical alteration. Consequently, there is a stratigraphic record of past atmospheres with a resolution and stability greater than in any other sedimentary environment on Earth (see Chapter 10).

The collaboration between scientists of many nations has resulted in the development and implementation of several conspicuous and suc-cessful research programmes. These programmes have materially advanced the study of ice and snow on planet Earth, and the recognition of inter-relationships with other elements of the global environment.

One outstanding glaciological achievement has been to draw attention to the role of ice in modulating global climate and in influencing oceanographic processes. Indeed as mankind's inadvertent interference with the climate becomes of real concern, so it becomes increasingly import-ant to assess the stability of ice sheets. It has been

postulated that a doubling of levels of carbon dioxide in the atmosphere during the next 50–100 years, with subsequent planetary warming (see Chapter 10), could destroy a significant fraction of Antarctic ice with the attendant catastrophic rise in global sea-level and flooding of inhabited coastal areas. This scenario for carbon dioxide, while clearly at the sensational end of the scientific spectrum, does illustrate the very close links between Antarctic ice, climate, oceans and the solid Earth. The West Antarctic ice sheet is likely to be most sensitive to warming, and so an inter-national programme is being developed to con-centrate on this region.

THE FROZEN BLANKET

Co-ordinated programmes over the past 30 years have enabled details of the Antarctic ice sheet to be elucidated and by only a handful of scientists. It is indeed the last part of our planet to be fully mapped and explored. Ground-based, airborne and satellite studies have shown that the Antarctic ice sheet is composed of three unequal parts. The

Icefall of the Ramsay Glacier in the Queen Maud Mountains

largest portion lies in East Antarctica (10.35 million km²), which reaches a maximum elevation of just over 4000 m at central Dome Argus. The Transantarctic Mountains (maximum elevation of 4528 m at Mount Kirkpatrick) effectively divide this ice mass from the smaller ice sheet in West Antarctica (1.97 million km²) on which the ice reaches a maximum elevation of 2300 m near Mount Woollard.

The West Antarctic ice sheet is flanked, in turn, by floating ice shelves in the embayments of the Ross (0.45 million km²) and Weddell (0.53 million km²) Seas. The Antarctic Peninsula comprises the third area (0.52 million km²) of complex glacierization with several small merging ice caps, ice shelves, extensive mountainous terrain, outlet glaciers, and ice-covered, offshore islands. Mapping provided a first step in understanding the processes by which the ice sheet responds to the influence of climate and sea level. Investigations of the dynamics and thermodynamics of large ice masses are central to the research of SCAR glaciologists because these forces may impinge

directly on fundamental questions of the stability of the ice sheet, and its fluctuations in size and shape on a variety of time-scales (decades, centuries, millennia and longer geological periods).

The ice is fed by an accumulation of snow which varies from a few centimetres per year in central regions far removed from oceanic moisture sources to several tens of centimetres to a metre in near-coastal localities. In normal circumstances, the ice, behaving as a viscous-plastic material, flows outwards from the continental interior under the influence of gravity. Measurements to date suggest that its speed is sufficient to balance the incoming snowfall (when averaged over many centuries). In this way, the ice sheet is able to maintain an approximately parabolic profile. The central regions are consequently very flat with gradients in the order of 1:1000. Towards the ice margin, surface slopes and velocities are higher, the ice is thinner and transmission of stresses from flow over the irregular sub-glacial bedrock makes the ice surface much rougher and undulating. Within 200–300 km of the coast, the

Erecting markers for measurements of ice movement on an outlet glacier in the Transantarctic mountains

Surveying glacier markers for studies of ice movement

Shear crevasses at the margins of a major Antarctic ice stream in Marie Byrd Land

ice may become channelled, either through peripheral mountains (as in the Transantarctic Mountains) where outlet glaciers develop, or, through fast-flowing zones within the ice sheet itself and termed ice streams. The latter features are found all around the continental margin but are best developed in Marie Byrd Land.

SHELVES OF ICE

Ice discharging from the continent flows into the sea. In many places it floats and continues to move outwards in the form of ice tongues and coalescing ice shelves, which spread out under their own weight and that of further surface snowfall in the interior and pressure from behind. The inner parts of the Ross and Weddell Sea embayments are filled with such ice shelves which attenuate in thickness from close to 1000 m near the grounding line (the point where the ice comes afloat) to about 200 m or less at the ice front. SCAR scientists quickly realised the importance of ice shelves. Because the ice shelves float on the ocean they are sensitive to

Front of the Ross Ice Shelf

oceanic influences, and their low elevation makes them also vulnerable to climatic warming.

Studies in computer modelling show that small changes in the level of the sea and the poleward transport of heat in the oceans and atmosphere may cause rapid and large-scale variations in the extent and configuration of the ice shelves. Such changes could remain undetected in the main body of the ice sheet over many hundreds of years.

To gain a better insight into ice shelves and contiguous parts of the ice sheet, SCAR discussed and endorsed a number of multinational, interdisciplinary research programmes—Ross Ice Shelf Project, which ran through the mid-late 1970s, the current West Antarctic Ice Sheet studies and the Filchner/Ronne Ice Shelf Project. Fundamental information is emerging that relates to the factors controlling the mass that is added to, or lost from, the ice shelves.

At the bottom of ice shelves, the ice is in direct contact with sea water. Depending upon the temperature contrast between water and the ice, the ice shelf base may melt and thin, or thicken by the accretion of small ice crystals to its underside. In 1978, glaciologists found that some inner portions of the Ross Ice Shelf accrete ice at the very slow rate of 2 cm per year, however, the region towards the ice front is characterized by melting at a rate of up to a metre or more per year.·

The dimensions of the ice shelves are in part controlled by the geographical configuration of the coastal zone in which they develop. Submarine banks or small islands often act as 'pinning' points and hold in the ice shelf. If such conditions were not met, an ice shelf would be split into large icebergs, helped by ocean currents, tide and wave action. Breakup still occurs but only at the leading edge of the shelf where spreading fractures the ice shelf to produce large tabular icebergs.

Icebergs from the seaward end of ice shelves and glaciers float out into the Southern Ocean. Estimates of the annual rate of production of icebergs still have considerable uncertainty, but values could be in the order of 2.3×10^{15} kg. The annual net total accumulation of snow in Antarctica is about 2.01×10^{15} kg. This suggests the mass budget for the whole ice sheet could be slightly negative, or just in balance. Other recent studies using observed rates of flow of ice suggest that the mass budget is up to 10 per cent positive.

ICE-OCEAN INTERFACE

The study of sea ice has assumed increasing importance to scientists during the past 5 years. SCAR recognizes the relevance of investigating sea ice in the Southern Ocean with reference to oceanography, climate and the special biological productivity of Antarctic seas. It has recently established a new Specialist Group on Sea Ice. What has been particularly attractive to international science is the transfer of technology and

Icebergs off Terre Adélie

As the sea starts to freeze it forms into ice pancakes

scientific skills developed in the Arctic to understand the practical role of sea ice in the Antarctic. Work, particularly from satellite imagery, has revealed the details on Antarctic sea ice. At its maximum, in late winter-early summer, Antarctic sea ice extends over an area of some 20 million km^2. Unconfined by surrounding land (in contrast to the geographical setting of the central Arctic Ocean) the sea ice is continuously broken and transported by the action of waves and currents. It is reduced in summer to something in the order of 3–4 million km^2. Most of the Antarctic sea ice, therefore, lasts only one year. The few regions in which the floes are more than first-year include the western Weddell Sea, the Bellingshausen Sea, and the Amundsen Sea. The presence of ice in the ocean has the effect of inhibiting the exchanges of energy, mass and momentum between the sea and the atmosphere. These factors play a crucial part in modulating global climate. The melting of the sea ice in the summer releases considerable fresh water to the near-surface layers of the Southern Ocean. In winter, salinity is increased by fractionation as the surface freezes over, and later as brine drains downwards out of the developing sea ice canopy.

Because of the inhospitable Antarctic environment, it has proved difficult and unattractive to study the basic processes operating in the Antarctic pack ice especially during its growth in the winter. This has placed particular importance on satellite observations which yield all-weather, day-and-night coverage. The extent and distribution of Antarctic sea ice has been detailed using the results of the electrically scanning microwave radiometer flown aboard the Nimbus 5 Satellite (see p. 42). In the future, it is anticipated that a number of SCAR nations will collaborate in a 10-year programme, involving ship-based projects and new satellite missions to investigate the interrelated glaciology, oceanography, meteorology and biology of the ice zone of the Antarctic sea. In this regard, the Antarctic has received less attention than the Arctic, where strategic and economic factors have stimulated much research on sea ice over the past two decades. The scene is set, however, for a major scientific advance on the southern sea ice during the next decade.

A FEEL FOR THE THICKNESS

Data on the thickness and volume of the ice sheet are fundamental to glaciological research in Antarctica. In the 1950s and 1960s, scientists traversed the ice sheet with over-snow vehicles undertaking seismic exploration to determine the thickness of the ice and the land structure below it—a technique refined by extensive use by the petroleum industry. Such work was slow, taking several months to complete only a handful of ice thickness records, and even these could be open to gross misinterpretation. Nevertheless, a sketchy and tantalising image of the ice sheet began to emerge. There was a clear demand for improved methods to provide rapid, accurate, and, moreover, continuous measurements of ice thickness to match the growing curiosity of glaciologists, meteorologists and geologists, and to feed increasingly sophisticated computer models of the dynamics of the ice sheet.

The technique of radio-echo sounding (RES) has revolutionized our knowledge of large ice masses. Many of the principles of the techniques are similar to those used in marine (acoustic) echo sounding. The system uses a downward-looking, very-high-frequency (VHF), pulse-modulated radar to probe the ice. Although scientists experimented with radar altimeters in the 1950s, it was not until glaciologists and radio engineers came together in the early 1960s with a common goal that RES became a reality. With the collaboration of scientists from many SCAR nations, more than half of continental Antarctica (7 million km²) has now been sounded by airborne radar on a 50–100 km grid, and more than 0.5 million km of flight track. The results of this programme have allowed the thickness of ice and the shape of the land surface beneath to be determined in substantial detail. The volume of ice is estimated to be 30 million km³.

Surprisingly, the thickest ice is located only some 400 km from the coast in Terre Adélie where a huge sub-glacial trench is filled with 4750 m of ice. Some of the thinnest ice, again paradoxically, occurs in the centre of the continent where a major sub-glacial mountain chain rises to a maximum elevation of 3500 m. In West Antarctica, the lowest point of the continent is located, some 2538 m below sea level, in the Bentley sub-glacial trench.

LAKES AND LAYERS

One of the more remarkable results of RES has been the discovery of substantial bodies of water, called 'lakes', beneath the ice sheet. These lakes have been detected from persistent, very strong bottom echoes which exhibit marked lateral continuity. They usually occupy hollows in the sub-glacial land surface under thick ice where temperatures at the bottom are expected to reach the pressure melting point. Many tens of lakes more than 5 km wide have been identified, principally in East Antarctica. The largest, near the Soviet Vostok Station, is about 8000 km². It is intriguing to speculate whether such lakes have any biological significance. Although buried under ice that exerts a normal pressure of 40 million pascals, their waters may possess considerable dissolved gases—from melting out of air entrapped in the ice and minerals dissolved from rock finely ground by glacial action.

Another detail to emerge from the RES work is that there are many internal reflecting layers in

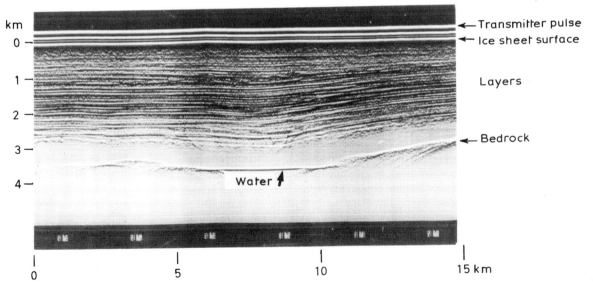

Typical record of radio-echo sounding through the ice sheet in East Antarctica. Clearly revealed are the bedrock, layers in the ice and a small sub-glacial lake

the ice. The detection of layers on radio echograms is due to reflections generated by small density changes in the upper 500 m of the ice sheet (due, for instance, to seasonal variations in snowfall and temperature). Below this depth, where ice reaches a constant density of about 900 kg per m^3, reflections are caused by acid impurities. These stem from volcanic activity elsewhere and have been deposited by snowfalls. It is possible to use the number, spacing, prominence, depth and deformation of such layers to infer a host of environmental processes. For example, estimates of the acidity of the ice can be made from the strength of the radar reflections. As depth in the ice sheet is equivalent to increasing age (the snow and ice form a stratigraphic succession), the layer variations in the strength of reflection provide data on climate, in particular on past stratospheric acid levels that are related to explosive volcanism over a period that extends back over thousands of years.

LOADING THE CRUST

The Antarctic crustal rocks are loaded with 2.7 × 10^{16} tonnes mass of the ice sheet and this depresses the true surface of the continent. The pattern of such loading is complicated, however, by the rigidity or stiffness of the Earth's outer shell. Using a computer, it has been possible to calculate from data on ice-thickness the amount of depression caused by the ice sheet. The amount of crustal sinking can be represented graphically in three-dimensions based on the assumption that the crust is currently in equilibrium as suggested by regional gravity anomalies. The greatest depression (950 m) is experienced in central East Antarctica. By restoring this pattern of warping to the current bedrock relief a new and 'adjusted' rock surface contour map may be produced. This 'adjusted' surface will approximate the shape of Antarctica prior to glaciation, and is the one that should be used for any comparison of land surfaces between adjoining fragments of the former 'super-continent' of Gondwana (see Chapter 6).

MORE REMOTE SENSING

Remote sensing from aircraft, and particularly satellites, offers scientists working in polar regions an unparalleled opportunity to gather synoptic data on a continental scale in a region of extreme cold, with long periods of darkness, often persistent cloud cover and where logistic support is complex and costly. SCAR scientists have been quick to use satellite results to aid mapping of the continent, to determine better the characteristics of the ice sheet and sea ice, and to provide the opportunity to monitor future changes in its mass balance. Both imaging and non-imaging systems have been used including the LANDSAT, NOAA and TIROS series. The processing of images from these missions, particularly from multi-spectral sensors, has enabled considerable detail of the ice sheet to be revealed—flow lines, zones of bare ('blue') ice, spatial changes in accumulation patterns, ice front positions, as well as inferences about underlying bedrock and flow regimes.

Satellite radar altimetry, which offers a precision of measurements to a few tens of cm, is providing a rich field of research for SCAR glaciologists where Antarctic studies are closely tied to other global research plans, particularly in oceanography. Data from the short-lived

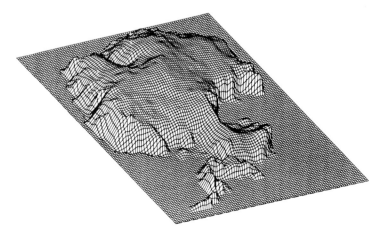

The surface of the Antarctic ice sheet plotted on a 50-km grid. Note the major indentations of glaciers and the ice shelves in the Ross and Weddell seas

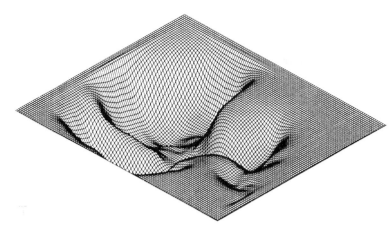

Antarctica plotted on 50-km grid showing the extent of the depression of the surface rocks as a result of the present ice load

SEASAT mission have already proved invaluable in setting base lines for the ice sheet—references required for measuring future changes in mass balance of the ice.

Microwave radiometers flown aboard satellites allow not only the gathering of information on the near-surface conditions of the ice sheet but also broad mapping exercises. The emissivity of ice and snow surfaces is related to both crystal size and the mean annual temperature. The data have been used to estimate variations of accumulation of snow over the Antarctic ice sheet and to indicate patterns of near-surface temperatures.

ARCHIVES OF THE PAST

SCAR scientists have dramatically advanced our knowledge of the internal physical and chemical composition of ice sheets and, as we shall see in Chapter 10, of global climate from deep drilling and ice core analysis.

Studies on ice cores drilled from the ice sheet enable us to learn about both past climatic conditions and simultaneous changes in the composition of the atmosphere which may reflect an underlying cause. Many nations are involved in parallel studies in Greenland and Antarctica because it is vital to establish the global extent of recorded changes and to determine whether or not they occurred at the same time.

Of global significance is the study of ice cores for the assessment of mankind's impact on the environment. Snow and ice in the Antarctic provide a unique storage facility for chemical impurities from the atmosphere—both particulate (such as volcanic ash) and dissolved. A study of

these products in an ice sheet provides information with background concentrations of major elements and a way of monitoring for levels of pollution. The use of lead tetraethyl additives in motor fuels, combustion of fossil fuels and large-scale industrial processes has been traced in Greenland ice by significant increases (up to 300 per cent) in lead, zinc and sulphur since the early part of this century. Surprisingly, however, no major changes are shown in cadmium, selenium, mercury and vanadium, which indicates that mankind's contribution to these products does not exceed natural levels. Organic pollutants (for example, pesticides such as DDT and dieldrin, and polychlorinated biphenyls) have also been detected in polar snow.

Contamination caused by fall-out of thermonuclear debris has been used to provide a new and very accurate means of determining accumulation in Greenland and Antarctica. By measuring gross beta-activity, or artificial tritium content, specific time reference levels can be established. Artificial radioactivity in Antarctic snow and firn was extremely low until the early 1950s. The first major increase occurred after the nuclear tests in the atmosphere over Bikini in 1954. A further marker horizon is at January 1965 (1963 in Greenland) related to the Northern Hemisphere atomic explosions of 1962. Using these reference layers, SCAR scientists have determined the rates of accumulation of snow for many sites. One interesting conclusion is that, in both the central parts of East Antarctica and the Ross Ice Shelf, accumulation is much less than was previously estimated.

8

Cold waters run deep

Knowledge of the movement of Antarctic water is vital to an understanding of global oceanic circulation and the world-wide balance of heat between the seas and the atmosphere. The Southern Ocean that surrounds Antarctica provides the only major linkage between the world's three great oceans: the Atlantic, the Pacific, and the Indian. The cold Antarctic waters flow northwards and because they are rich in oxygen, their circulation results in significant aeration of the world's oceans.

Despite their very low temperatures, the seas around Antarctica are productive. The surface temperature of the Southern Ocean ranges from about 4°C in the warmer northern waters to around −1.8°C in the areas close to the Antarctic continent (see Fig. 8.1). In winter, vast areas of the Southern Ocean are covered by ice. With the onset of spring, light returns, most of the sea ice breaks up and dissolves, and tiny plants of the phytoplankton start to multiply. In inshore areas the algal bloom can be so intense that it colours

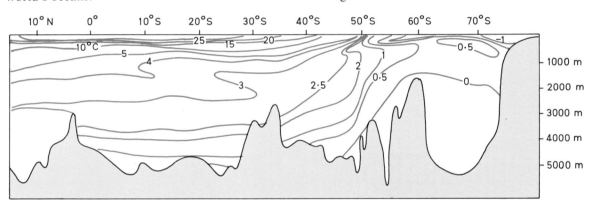

Fig 8.1 Distribution of ocean temperature in a vertical section along the meridian 30° West

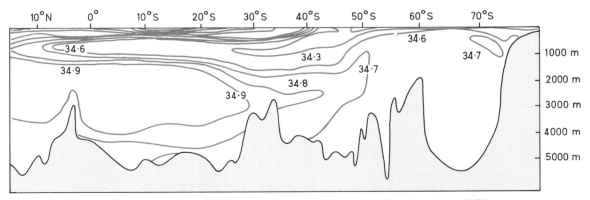

Fig 8.2 Distribution of salinity, in parts per thousand, in a vertical section along 30° West

the sea. This burst of plant growth in spring provides the food for a vast and complex web of other organisms (see Chapter 12).

Throughout most of the oceans, water masses hover at their respective density levels, circulating horizontally but mixing upward and downward with imperceptible slowness. The Southern Ocean, however, is a crucial exception. Vertical circulation is an important component, caused by the winter freezing of a vast area (about $20 \times 10^6 \, km^2$) to produce a denser, high salinity residue which sinks, generating the Antarctic Bottom Water (ABW).

This sinking—called the 'thermohaline circulation' because it is driven by differences in water temperature and salinity—powers the deep movement of water toward the Equator (Fig. 8.2). In the depths of the Antarctic seas, cold, salty water sinking from the surface displaces the bottom water, squeezing it along levels of constant water density (Fig. 8.3), where its lateral movement is modified by the shape of the ocean

Net sampling on fast ice, north of Syowa Station

bottom and deflected by the rotation of the Earth. The greatest single source of bottom water is the Weddell Sea. There, harsh weather, a high production of ice, and peculiarities of surface flow create some of the densest ocean water in the world, highly saline and below the freezing point of fresh water.

The high salinity of winter surface waters of the Weddell Sea is a consequence of the formation of sea ice. As sea water freezes, salt is expelled from the crystalline structure of the ice; the salinity of pack ice is much lower than that of seawater and the remaining salt is concentrated in pockets of liquid brine within the ice. As the ice ages, this brine drains, increasing the salinity and density of the sea water beneath. Salinity of the ocean reaches a peak in winter, after the formation of sea ice, then is reduced in summer as the ice pack breaks up and melts. The prevailing winds drive the melting floes out to sea.

Dense, cold (katabatic) winds, sinking from the Antarctic interior, chill the shelf water, reducing

Beam trawling, north of Syowa Station

Bringing aboard the beam trawl

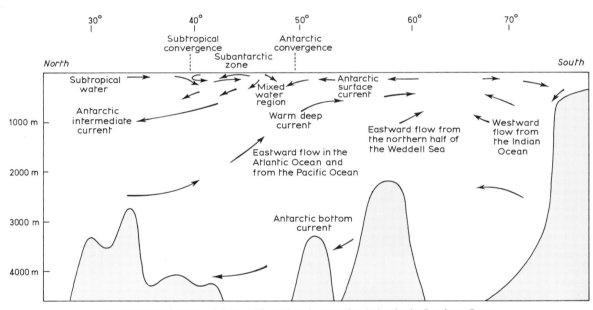

Fig 8.3 Main features of the two-dimensional ocean circulation in the Southern Ocean

its temperature to near freezing. The result is that the deeper, denser water sinks down the continental shelf into the deep sea. There it joins the deep circulation that channels it towards the Equator, and even into the Northern Hemisphere, thus forming the layer of Antarctic Bottom Water that spreads over the depths of the world's oceans.

As the dense, high-salinity water sinks into the depths, deep-seated warmer water of a more northerly source wells up to replace it. And farther from the Antarctic continent, at a latitude of about 50° South, another shift in the vertical hierarchy occurs (see Fig. 8.3). The surface water there has a lower salinity because of heavy snowfalls and melting pack ice. Driven northward by prevailing winds, this water encounters saltier and considerably warmer water moving southward from temperate latitudes. This meeting of water masses occurs at the Antarctic Convergence (see Fig. 8.4).

The Antarctic surface water is close to the freez-

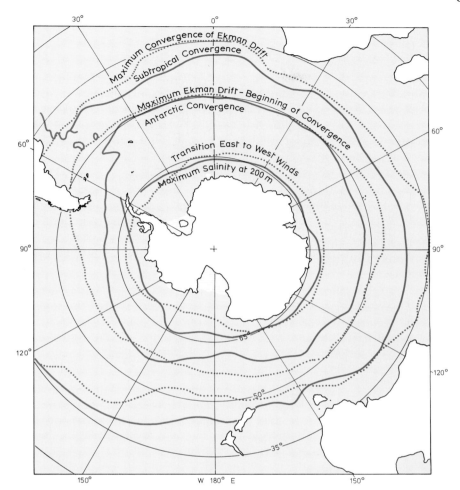

Fig 8.4 Wind and water boundaries in the circumpolar ocean

ing point, and therefore is more dense than the temperate water masses, which may be as warm as 15°C. Thus the northbound water sinks beneath the warmer masses at the zone of convergence and continues its northward progress at a depth of about 900 m. This, the layer of Antarctic intermediate water, is characterized by its low salinity.

The Antarctic Bottom Water makes a much longer journey. Some of it circulates eastward around Antarctica. But much of it flows directly northward into the western South Atlantic. Concentrated in a narrow boundary, it hugs the South American coast as far as the Equator, then, for reasons still not understood, moves eastward and follows the mountainous spine at the centre of the Atlantic, the Mid-Atlantic Ridge. By the time it reaches mid-Caribbean latitudes, the bottom

current is so mixed with overlying warmer water layers that it is difficult to identify.

*   *   *   *   *   *   *

The explorations of the Antarctic seas in the early 19th century were quickly followed by exploitation of the seals and whales, because in contrast to the Antarctic continent, the Southern Ocean is relatively productive biologically. It was this richness of life that has provided the spur for many investigations of the cold seas.

The expedition of the German ship *Gauss* during 1901–3 was the first to describe the sinking, between latitudes 40° and 60° South, of cold surface waters from the Antarctic. Ten years later, during the long drift of the ice-bound *Deutschland*, scientists aboard determined the structure and circulation of the waters in the Weddell Sea and explained for the first time the formation of the deep, cold 'Antarctic Bottom Water'. However, the expansion of the whaling industry in the 1920s and the rapid demise of the principal whale species led to the first systematic study of the Antarctic circumpolar ocean as a whole. The *Discovery* investigations confirmed:

> That the typical three-layer structure of cold surface Antarctic water, relatively warm deep water, and cold bottom water extended all around the continent;

> That the rather rapid transition between the characteristic Antarctic and subantarctic water, the Antarctic Convergence or 'polar front', could be traced all around the world, as could a somewhat less well defined discontinuity further north, the subtropical convergence.

Over the past 30 years, scientists working in international collaboration on the properties, populations and processes of the Antarctic circumpolar ocean have used a range of new techniques to improve their understanding of this vast, inhospitable and little-travelled region. The incentives for such collaboration have been essentially academic. For example, the contrast between the Northern Hemisphere oceans, with their meridional boundaries, and the Antarctic circumpolar ocean, which is nearly zonal, provides a useful check on ideas on ocean circulation. Certainly the relation of the physical conditions to the complicated food webs must be understood if mankind is to control any harvesting of the vast quantities of Antarctic krill—the most obvious marine resource that can significantly increase our harvest of protein from the sea. Above all, though, oceanographers and climatologists now recognize increasingly the importance of the Southern Ocean, its ice, and its currents in the maintenance of climate. That ocean is fundamental to climatic change because its waters store and transport heat, salt, and dissolved gases such as carbon dioxide: these components effect the density and mobility of the water.

So there are good reasons for international studies of the Antarctic circumpolar ocean. The programmes of the IGY included an extensive Southern Ocean component, although most of the hydrographic observations were made on supply routes to continental bases. Antarctic supply ships continue to collect valuable data, extending the important, if incomplete, series of oceanic measurements taken in successive periods of time. A particularly important and extensive data set was collected by the USNS *Eltanin* between 1962 and 1972 on 52 cruises covering more than 660 000 km of track. Scientists from many countries collected physical, chemical, geological and biological information from most of the Southern Ocean south of 35° South. From 1974 to 1979 the *Eltanin*, renamed *Islas Orcadas* and operated from Argentina, continued the circumpolar survey into the South Atlantic, and in 1974 part of the southwest sector of the Indian Ocean was surveyed by USNS *Robert D. Conrad*. So, studies of the Antarctic circumpolar ocean, between 1972 and 1979, have provided observations basic to our understanding of the circulation. The *Eltanin/Islas Orcadas* data are a primary source for the *Southern Ocean Atlas* (Columbia University Press, 1982), which illustrates our improved knowledge of the anatomy of the ocean—the thermohaline structure.

Oceanographers are increasingly studying the way in which the various masses of water are formed, especially that of the Antarctic Bottom Water, the densest water in the open ocean. Estimates suggest that the total northward flow, all round the circumpolar ocean, is about 34 million tonnes per second, providing the major source of the cold water in ocean depths far to the North. Precise determinations of temperature and salinity from modern instruments are making it possible to differentiate between bottom water found in the Weddell Sea and in other locations such as the Ross Sea. The International Weddell Sea Oceanographic Expedition (IWSOE) paid particular attention to the Weddell Sea (the primary source of bottom water). The expedition made extensive use of ice-breakers in order to provide oceanographic data within the summer ice fields. But in general, IWSOE scientists studied only areas free from ice until October–November 1981. Then a joint US–USSR programme from the Soviet vessel *Mikhail Somov* undertook observations well within the sea-ice cover. Such work led to a better understanding of formation and modification of water masses.

Nor have the ocean currents been neglected. The winds are predominantly from the West, driving what was called the West Wind Drift but what is now referred to as the Antarctic Circumpolar Current. The total transport of this

Deploying a CTD sonde for measuring water conductivity, temperature and depth

current is calculated to be about 130 million tonnes per second (several times larger than the Gulf Stream). The North-South motions indicated in Fig. 8.3 are superimposed on a much stronger zonal flow. During the 1970s, scientists working on the International Southern Ocean Studies (ISOS) concentrated on measuring the currents in the Drake Passage and in a region south-west of New Zealand. They used current meters, thermistor chains, and bottom-pressure gauge moorings, as well as the more conventional methods to determine the temperature and salinity in relation to depth. They found that the structure of the Antarctic Circumpolar Current was complicated, with multi-filament structures especially near the frontal zones of the Antarctic Convergence. Near the Drake Passage, cores of strong currents about 50 km wide were separated by 150-km regions where the currents were slack. As in the Northern Hemisphere Ocean, there were eddies in the current, with typical time-scales of 14 days and distance-scales of about 40 km. Oceanographers have detected this kind of eddy variability in many other regions of the Southern Ocean.

In the Drake Passage, satellite images of the sea surface revealed that eddies in the current at 2700 m depth were related to meanders in the Antarctic Convergence. During the ICSU-WMO First GARP Global Experiment (FGGE) of 1978–79, surface currents scale were studied incidentally from the varying locations of drifting buoys. The buoys constituted one component of the FGGE. They measured sea surface temperature and

Oceanographic and glaciological research near the Riiser-Larsen Ice Shelf, Dronning Maud Land

Local hydrographic survey close in to an ice shelf

Observation by satellite is of particular importance to the study of such a large, remote and inhospitable area as the Antarctic circumpolar ocean. By means of passive microwave images, it is possible to map the advance and retreat of sea ice—a major force that affects variations in climate, on seasonal and interannual time-scales. Visible and infra-red images help to define the variability in regions free from cloud.

A number of other techniques used in the brief SEASAT mission of 1978 hold great promise. Average sea levels can vary from $+6$ to $-6\,m$ depending on whether they are over mid-ocean ridges or deep trenches. Such measurements can be gained quickly and with great accuracy by altimeters carried on satellites. The data make it possible to estimate the mean 'slope' of the sea surface, and this can be related to ocean currents. Scatterometers provide measurements of speed and direction of near-surface winds.

In no area is international co-operation more important than in satellite studies. Not only is co-operation essential in the design and the launching of satellites but also in their calibration and comparison with surface observations. International agreement is required on the standardization and transmission of the immense amount of data involved in satellite studies. The challenge and importance of the Antarctic circumpolar ocean are now being met by a host of modern developments in technology. The improved understanding of the Southern Ocean and its system will be reflected in increasingly realistic computer models, developed and interpreted by the international community of marine science. A realistic global model of a coupled atmosphere and ocean, with ice, is now a conceivable goal of international co-operation.

surface air pressure and transmitted their observations several times a day, to an orbiting satellite, which located the position of the buoys (see also Chapter 9). From the tracks followed, currents were deduced. Similar satellite-location methods have been used to track transponders on icebergs.

9

Antarctic meteorology

Historically, Antarctic meteorology has progressed through three distinct phases. Prior to the IGY, it was mainly a matter of describing basic climatological features. Just before the IGY, it included establishing the first permanent meteorological stations on the coasts of the continent itself—the Antarctic Peninsula and some of the subantarctic islands already had well-established stations. Notably in this period, the Norwegian-British-Swedish (NBS) Antarctic expedition to 'Maudheim' carried out some primary meteorological research into solar radiation and the energetics of the boundary layer over a snow surface, and studied the troposphere.

The second phase began with the IGY and continued until the early 1970s. The main developments came with the understanding of the Antarctic weather systems and with further developments in physical meteorology of the Antarctic. For the IGY, an 'International Antarctic

Weather Central' was established at the US Little America Station in the Bay of Whales at the head of the Ross Sea. This co-operative enterprise could be said to have 'discovered' how the Antarctic meteorological system behaved. American and Soviet meteorologists defined in some detail the circulation patterns in the troposphere (3–10 km up) and lower stratosphere (20–30 km up; see p. 97). The work of the analysis centre was later continued in Melbourne, Australia, at an International Antarctic Analysis Centre that was maintained for a number of years. In 1966, the centre was relieved of routine data and chart analysis by the establishment of a World Weather Watch Southern Hemisphere Analysis Centre (also in Melbourne). The analyses focused attention on the spring warming of the stratosphere. Such warmings occur much later in the Southern Hemisphere than in the Northern Hemisphere, and invariably start in the Indian Ocean sector. In

Dawn and a frontal system approaches the Antarctic Peninsula

South Africa, the Weather Bureau published a series of synoptic analyses and climatological studies of the Southern Hemisphere.

The IGY data facilitated the initial estimates of the mass of air moving to and from high latitudes ('meridional mass transport'). Vertical motion in the troposphere was related to both the outward katabatic flow of cold air from the surface of the continent, and to the intense westerly winds in the winter stratosphere, usually referred to as the 'polar vortex'. None of these discoveries would have been possible without the ring of surface and upper-air meteorological stations around the continent and on the polar plateau.

COMMUNICATIONS AND METEOROLOGICAL SATELLITES

In its third stage, since the mid-1970s, Antarctic meteorology has become an integral part of the World Weather Watch, absorbed into the fabric of global numerical analysis for weather forecasting and for models of global climate. The interest in Antarctic meteorological weather systems has spread beyond the bounds of SCAR scientists into

Drifting snow blowing off the ice shelf in Halley Bay

Servicing an automatic weather station on the Antarctic Plateau

the major meteorological numerical analysis and forecast centres (for example, the European Centre for Medium Range Weather Forecasting). This interest has necessitated improvements in methods of communication between Antarctica and the Global Telecommunication System (GTS). Geostationary communications satellites are being used as part of the GTS because the early receipt by the numerical centres of the surface meteorological observations and measurements of pressure, temperature, humidity and winds in the upper atmosphere is vital to the accuracy of the analyses.

In areas such as the Antarctic and the Southern Ocean where there are few meteorological stations, the development of meteorological satellites has proved a boon. Pictures taken by satellites in the visible and infra-red parts of the spectrum have enabled meteorologists to position weather systems correctly. This in turn has shown the principal areas where depressions (cyclones)

Blowing snow at Halley Station

form, and the tracks which they take as they move around and into the continent. The region of the lowest pressure around the continent, known as the circumpolar trough, varies its latitude by a few degrees over a period of a few years. This suggests that there are interannual variations in the circulation patterns linking the Antarctic to the temperate and subtropical latitudes across the Southern Ocean.

SEASONAL VARIATIONS AND TRACE MATERIALS

Regular measurements of atmospheric turbidity and estimates of total ozone revealed that a veil of fine stratospheric dust from the Mount Agung volcanic eruption, in Bali, in March 1963, could not penetrate the core of the winter polar vortex until the stratosphere was sufficiently warmed, in late spring, by solar radiation. This warming effects a slackening and reversal of the strong poleward temperature gradient that drives the stratospheric circulation in winter.

More recently, monitoring ozone has produced an even more dramatic result. At Halley, the spring ozone amounts, isolated from mid-latitudes within the polar vortex, are 30 per cent less than a decade ago. This is ascribed to the catalytic destruction of stratospheric ozone due to chlorine. The source of this chlorine is probably man-made chloro-fluorocarbons (freons), used as refrigerants and aerosol-can propellants. Ozone amounts recorded at Syowa and by instruments aboard American satellites have confirmed the Halley results. Taken together with the results of a balloon-borne ozone-sonde campaign carried out in 1985 by East German scientists at Novolazarevskaya, it is concluded that the 'ozone hole' occurs over half of Antarctica and above the Weddell Sea in spring. Further studies are being planned to investigate this important phenomenon.

SCAR scientists are investigating the seasonal variation in amounts of minute traces of natural salts, stable isotopes, gas, and pollutants (frequently referred to collectively as 'trace species') that reach Antarctica. Just north of 60° South latitude, an American team released methane labelled with deuterium, a heavy isotope of hydrogen. The team then analysed air samples taken at stations all around the continent and at the South Pole to find where and when the tracer reached the continent.

AUTOMATIC WEATHER STATIONS

Improved sensors on new meteorological satellites will provide more data world-wide, including Antarctica, on a diversity of parameters, such as winds, and will even make some observation through cloud. But even these advanced satellite systems will require verification of measurements by observations from sea, land, and ice-surface stations.

During the First GARP Global Experiment (FGGE), organized jointly by ICSU and WMO in 1978/79, the Antarctic station network was greatly augmented. Many buoys were released into the oceans of the Southern Hemisphere and their positions and the data which they gathered relayed back by polar-orbiting satellites. The buoys provided data on sea-level atmospheric pressure and sea-surface temperature for vast areas which hitherto were without even ships' reports. The impact on synoptic analyses was immediate. The pressures in depressions (cyclones) over the oceans were found to be even lower than previously thought.

The network of buoys in the FGGE was designed to provide wide meteorological coverage for that experiment. No comparable network has since been established although drifting buoys continue to be used extensively to study specific problems in limited areas of the oceans of the world. It is likely, however, that a new network will be deployed in the Antarctic for the World Climate Research Programme (see Chapter 10) and the projected study of the sea ice zone.

In recent years, scientists have deployed a number of automatic weather stations (AWSs) on the polar plateau, coastal East Antarctica and on the east side of the Antarctic Peninsula. The AWSs report surface pressure, temperature and wind via polar-orbiting satellites. Such a station at Byrd has now been in continuous operation for over three years. Originally intended for climate studies of the cold katabatic winds draining off the ice sheet, the AWSs are now providing invaluable data for numerical analyses.

MODELLING THE WEATHER AND CLIMATE

Modern techniques are making it possible to investigate the interaction between the Antarctic weather systems and those in temperate latitudes. Improved accuracy of weather forecasts, up to five days ahead over large areas of the Southern Hemisphere, is possible now that data can be gathered from all levels of the atmosphere. The Antarctic is no longer the isolated 'white hole' that it was long considered to be. Now it is an integral part of all models of the Earth's climate system.

Numerical analyses on both synoptic and climatological time-scales will continue to be of importance, and the demand for good meteorological and climatological data from the Antarctic and the Southern Ocean will carry on growing.

Band of low stratus lies along the Antarctic Peninsula

Weather station for a summer field camp in the Prince Charles Mountains

Patches of sea fog and an iceberg

Parhelia, or mock suns,

The currents of the Southern Ocean and neighbouring seas and sea ice cover will be major focuses of attention (see Chapters 7 and 8). The number of automatic weather stations deployed on the mainland is expected to be increased, first for studies of the dynamics of katabatic winds, and second to fill large gaps in the station network that exist or arise because a manned station has been closed.

ENVIRONMENTAL DAMAGE AND POLLUTION

The world-wide concern for environmental matters also enhances the need for reliable monitoring of turbidity, carbon dioxide and trace species and, most importantly, stratospheric ozone. The depletion of ozone increases the risk of undue exposure to ultraviolet light. The risk is as yet but slight, but the world-wide use of

Sunset and brash ice

chlorofluoro-carbons as propellants for aerosols, and the production of oxides of nitrogen in engines, are now thought to be playing an important part in depleting ozone.

In remote sensing, the next generation of satellites presents new opportunities for improving measurements of surface and atmospheric temperatures, for determining the thickness of sea ice as well as the standard areal coverage, and for stratospheric studies. Ground-based laser radars ('lidars') and stratospheric–tropospheric (ST) radars are expected to be developed and used for continuous measurement of tropospheric and stratospheric temperatures and winds.

Meteorology and atmospheric chemistry are directly concerned with detecting atmospheric pollution. The Antarctic is a 'clean' location and gives the threshold value of the global atmosphere at any one time. Increases in pollution monitored in the Antarctic can provide an early warning of impending danger to the rest of the world. Therefore, it would be as well to keep the Antarctic 'clean' and to ensure that no manufacturing industrial activity takes place within it.

By the turn of the century, many of the research projects and programmes outlined here will have been completed. New questions will undoubtedly arise from the work done. Throughout the period and well into the next century, meteorological observations and atmospheric soundings will be needed both for global weather forecasting and for local and climatological monitoring. Air travel to and from Antarctica, both for science and tourism, is likely to increase and will underscore the need for more weather forecasting in the region.

10

Cold climates and frozen atmospheres

Because of the curvature of the Earth, and because the axis around which the Earth rotates daily is inclined, by 23.3°, to the plane of the planet's annual path around the Sun, the net annual radiation budget in the polar regions is less than elsewhere in the world. Most of the solar radiation that the Antarctic continent does receive is reflected out into space by the mirror-like ice sheet. Antarctica is a net loser of heat by radiation, and the continent is *the* major global heat sink and this has a widespread effect. The difference between the solar radiation received in the tropics and the radiation lost to space from the polar regions establishes energy transfers that drive the global atmospheric circulation. These energy transfers also create such features as the convergence zone, the circumpolar trough and high-pressure belts.

An understanding of the role of the Antarctic in the global climate system requires knowledge of the interaction of both sea ice and the ice sheet with the atmosphere and the oceans. The Antarctic interacts with the total climate system on a wide variety of time-scales.

The ice sheet is of importance on scales of tens of thousands of years or more, although possible instabilities in ice flow (particularly of the marine West Antarctic ice sheet) may be important on scales of hundreds to thousands of years.

The cold dense water produced as a result of ocean cooling by melt-ice from the ice shelves and icebergs, and by salt rejected from growing sea ice, sinks onto the continental shelf and thence becomes transferred to the deep ocean basins. These masses of Antarctic 'intermediate' and 'bottom' water (see Chapter 8), and the transport of heat by the Southern Ocean circulation influence climate on time-scales from decades to centuries. Anomalies in the extent of the sea ice, its concentration and distribution are important on not only annual but also decadal scales. Over the

Collecting samples of ice from Taylor Glacier, Southern Victoria Land

past 30 years, major advances in understanding of the role of the Antarctic in the global climate system have come from studies in a range of disciplines.

ATMOSPHERIC PROCESSES

Throughout the history of Antarctic exploration, weather observations have been made with care and devotion. The permanently occupied stations, established during and after the IGY, form part of the global meteorological network and provide a detailed climate record of Antarctic atmospheric conditions for the past 25 years or more. Most of these manned stations, however, are on the fringe of the continent with only a few on the interior ice sheet. Automatic weather stations (AWSs) are now being increasingly used. The station coverage of the continent and surrounding ocean has been developed in the past few years.

Within the limitations of the areal coverage and the length of the meteorological record, SCAR scientists have defined the broad features of the Antarctic atmospheric circulation and the role of the Antarctic as a 'heat sink' in the global general circulation. However, understanding of the climatic processes in the Southern Hemisphere remains rudimentary. Computer models of circulation and climate of the Antarctic perform poorly at present—probably because they do not allow adequately for the exchanges of heat, moisture and momentum between the atmosphere and underlying surface, or for the radiation reflected, absorbed, and re-radiated by high-latitude clouds.

Scientists have studied exchanges of energy between the surface of the ice sheet and the atmosphere at a number of points; however, they have not yet adequately determined the links between these and the medium- and large-scale energy balances and the cold winds (katabatic) that drain off the ice sheet. Simple models can reproduce some of the features of the katabatic flow but they are not yet realistic enough to throw any light on the complex interactive relationships between the topography of the ice sheet, the surface winds and the surface energy balance.

In recent years, observations of the constitution of the Antarctic atmosphere, its content of trace gases, and aerosols, have provided the beginning of a database both for monitoring pollution and to aid the interpretation of the record of gases and aerosols trapped in the ice sheet.

INTERACTION OF SEA ICE, OCEAN AND ATMOSPHERE

Sea ice is an important interactive and dynamic element in the global climate system. The cover of sea ice markedly changes the surface reflectiveness (albedo). Further, it insulates the ocean from the atmosphere and exchanges salt with the ocean. Hence it affects the radiant and turbulent exchanges of energy, the transfer of momentum between atmosphere and ocean, and the vertical mixing and salt balance in the ocean. Sea ice modifies the seasonal temperature cycle, delaying cooling of the atmosphere in autumn by the release of 'latent heat' as an increasing amount of water becomes ice, and having the reverse effect during ice-melt in spring. The movement of the ice is also important, particularly because the generally divergent flow around Antarctica modifies both the concentration of salt and the heat budget in the ocean by the net flow towards the equator of ice.

Since the early 1970s, satellite sensing, particularly using passive microwave radiometers, has provided data on the extent and concentration of Antarctic sea ice, the seasonal and interannual variation of the extent of the ice, and some information on the type of ice (new ice, year-old ice, or multi-year ice). The total extent of Antarctic sea ice of greater than 15 per cent concentration varies from about 3 million km^2 in February to more than 20 million km^2 in September/October, and—even at maximum ice extent—25 per cent of the area within the ice boundary is open water, drastically increasing the transfer of energy between ocean and atmosphere. There are large variations from one year to the next in the extent of the ice and also indications that, in most parts of the Southern Ocean, the sea ice retreated between 1973 and 1979.

The variability in the seasonal changes between different regions can be pronounced, with the amplitude of the annual cycle being greatest in the Ross and Weddell Seas. The strong vertical mixing in the ocean in these two regions means that they are the major areas in which Antarctic Bottom Water is formed. There are also year-to-year regional variations in the extent of sea ice, which may be expected to affect the structure of the oceanic and atmospheric circulation at high latitudes.

Although satellite sensing has provided data on the extent and variability of the ice, such advances are not matched by an increase in knowledge of the properties and characteristics of the ice within the pack. The interior of the Antarctic seasonal sea ice zone is the largest area on the Earth's surface as yet unexplored in all seasons. Comparatively little is known about the thickness distribution of sea ice, snow accumulation on the sea ice, the drift of ice, the flow of heat from the ocean to the ice, and other climatically vital parameters (see Chapter 7). A number of detailed studies of the processes of interaction between sea ice, ocean, and atmosphere have been made, but mostly at near-coastal sites where there is continuous cover of ice semi-permanently attached to the land ('fast ice'). The knowledge gained in these studies is not directly transferable to the ice of less than 100 per cent concentration within the bulk of the seasonal sea ice zone.

THE CLIMATE RECORD FROM ICE CORES

The deep ice of the inland ice sheet may be several hundreds of thousands years old. It contains fine particulate material and has a chemistry representative of conditions at the time of deposition. The continuous environmental record that can be obtained from examining Antarctic ice cores is arguably the most valuable resource of the continent.

Deep ice cores (greater than 2000 m) have been recovered from the research stations at Byrd and Vostok; numerous intermediate and shallow depth cores have been obtained from other sites. These cores reveal atmospheric environmental changes over the past 150 000 years or more—a period which spans the coldest part of the last ice age, through the last interglacial and into the previous ice age, the climatic transition, and the present warm climatic stage. It is the period of Earth's history critical in Man's emergence. Ice cores provide evidence of drastic environmental changes in the past.

Analyses of ice cores can reveal:

Atmospheric temperatures, recorded in the relative abundance of the heavy stable isotopes of oxygen and hydrogen;

Clean sampling of snow blocks for use in global pollution studies

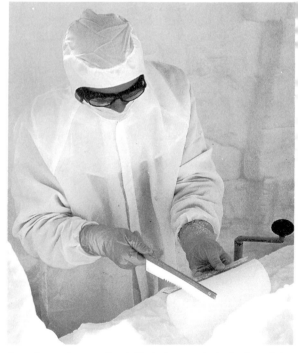

Sampling an ice core for stable isotope analysis. Such work is essential for unravelling the information that the ice holds about climates past

Searching for volcanic
horizons in an ice core

Atmospheric pressure, related to the height of the ice sheet surface when the gas was trapped in the ice;

Precipitation, estimated from snow accumulation which is determined by a variety of techniques, including the detection of annual layering in the stable isotopes and other variables;

Atmospheric loading of both volcanic and non-volcanic dust, the latter being related to the extent of deserts and the vigour of wind transport.

Chemical analysis of the ice cores gives evidence of the extent of sea ice (from marine impurities), volcanic activity, solar-terrestrial phenomena (from beryllium-10 and nitrates) and changes in the microfine particulate material associated with human activities. In particular, ice cores offer the main, if not perhaps the only, hope of establishing the history of the global variations of carbon dioxide, methane, and other climatically important trace gas concentrations prior to the start of continuous measurements in the 1950s.

ICE CORES

Ice cores provide a record of past environmental and climatic changes on timescales up to about 150 000 years. Most of the water in the oceans is made up of molecules containing the common isotope of oxygen, ^{16}O, but there is also some ^{18}O present. Molecules containing the heavier isotope (^{18}O) evaporate less easily, and the proportion of each of the two oxygen isotopes present in precipitation depends on the mean temperature of the cloud at the time. So the ratio of heavy to light isotopes present in old ice reflects the temperature at the time that ice was being laid down as snow. Data from the Vostok ice core therefore provide (Fig. 10.1) a direct measure of global temperature changes over the past 150 000 years, from a core more than 2000 metres long.

The important features of these changes are that climate has been up to 10°C colder than at present for most of the past 150 000 years, with large periodic or quasi-periodic variations in temperature both within the cold intervals (ice ages) and the shorter, warm intervals—the so called 'interglacials'.

Looking in more detail at the record of the past 30 000 years (Fig. 10.2), we can see the marked changes that occurred at the beginning of the present interglacial. Compared with the interglacial, the glacial maximum was not only 10°C colder but was also drier and experienced strong winds, as shown by the large numbers of dust particles carried to Antarctica and deposited in the ice. Direct measurements of the ice layers show that the accumulation of snow was only half as rapid as today, confirming the low level of pre-

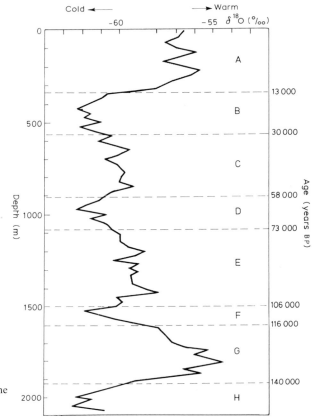

Fig 10.1 Climatic change over the past 150 000 years from the Vostok ice core. Lower $\delta^{18}O$ values characterize the colder climatic periods B, D, F, and H

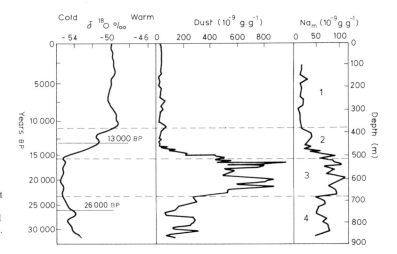

Fig 10.2 Smoothed stable isotope record, dust and marine sodium concentrations (Na_m) measured in the Dome C ice core and plotted as a function of age and depth (metres of ice). Horizontal lines delimit climatic stages

cipitation. But there is no evidence of the kind of dust particles and acidity to suggest that more volcanoes were active during the ice age.

CARBON DIOXIDE

Climatologists today are greatly concerned at the probability that the rapid increase in the concentration of carbon dioxide in the air, caused by human activities, may cause the world to warm significantly through the 'greenhouse effect' (see Fig. 10.5). Studies of bubbles of air trapped in Antarctic ice cores provide evidence of how the concentration of carbon dioxide varied naturally before the human influence became pronounced. During the latest glacial maximum, 20 000 years ago, the concentration of carbon dioxide (Fig. 10.3) was about 200 parts per million by volume (ppmv), compared with about 280 ppmv in the 19th century; direct measurements were first made only 100 years ago. At the beginning of the present interglacial, some 10 000 years ago, there was a

sharp change to a different regime, but the concentration never rose above 300 ppmv until the present century. Today, it is about 340 ppmv and rising rapidly—Fig. 10.4 shows the rapid rise from ice core data going back to the mid-18th century.

The low concentration of carbon dioxide during the ice age was probably a result of the coldness of the oceans, because cold water absorbs carbon dioxide more efficiently than does warm water. Nevertheless, it may have contributed to the stability of the glacial state through a reduced greenhouse effect.

Detailed studies of ice cores covering the past 100 years or so show traces of impurities associated with volcanic activity and nuclear tests, but no clear trends for temperature and snow accumulation. The baseline provided for climate modellers studying the possibilities of a greenhouse effect caused by carbon dioxide is probably the most immediately important piece of data to emerge from studies of these shorter stretches of core.

DYNAMICS OF THE ICE SHEET

Interpretation of the climatic records from ice cores and of temperature-depth profiles in ice sheets requires a detailed knowledge of ice dynamics. Only with this knowledge can the geographic origin and age of the deep ice be determined, and it has not been possible to do this accurately for many of the existing cores. Measurements in the boreholes and on the ice cores themselves provide useful data for the study of ice dynamics and for modelling the ice sheet.

Ice dynamics describe part of the hydrological cycle: ocean–atmosphere–ice sheets–ocean. In general, the time constants for significant changes

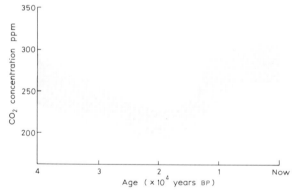

Fig 10.3 Estimated range of atmospheric carbon dioxide during the past 40 000 years

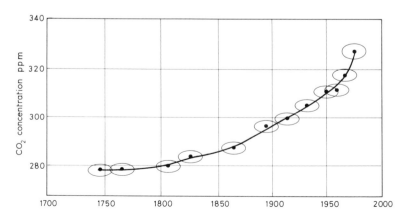

Fig 10.4 Measured mean concentration of carbon dioxide plotted against the estimated mean gas age

of the Antarctic ice sheet are long compared with atmospheric and oceanic processes, so ice sheets have a stabilizing influence on the cycle. However, possible instabilities in the flow of ice sheets could have drastic effects on both sea-level and climate on a time-scale, estimated from inadequate knowledge, from one to several centuries. Such instabilities could be due either to inherent characteristics of ice flow, or to triggering by external changes such as atmospheric temperature or precipitation.

Numerical models of the ice sheet have been developed using measured parameters as boundary conditions and the knowledge of ice dynamics obtained both from field and laboratory experiments. The computer models have been used to calculate various properties of the ice sheet, such as the temperature within the ice, flow regimes, and basal conditions (for example, melting). They have also been used to estimate the response of the ice sheets to environmental changes (such as climatic warming induced by carbon dioxide), and in modelling the climates of ice ages. Further progress is hindered by the limited nature of our knowledge of Antarctic ice dynamics. In particular, glaciologists have only rudimentary understanding of sliding processes at the base of the ice, and of melt and refreezing processes that occur at the base of ice shelves and at the seaward front of the inland ice.

RESEARCH PRIORITIES

The complexity of the interactions in the global climate system, particularly in the Antarctic, presents many new challenges. Antarctic climate research is actively continuing as a contribution to the World Climate Research Programme (WCRP), a global programme co-ordinated jointly by ICSU and the WMO, which aims to determine to what extent climate is predictable and to what extent Man influences climate.

The 1983 report of the SCAR Group of Specialists on Antarctic Climate Research identified the main deficiencies in understanding the data of the region, and recommended that priority should be given to a number of research topics. Much of this work is being incorporated into national programmes on Antarctic climate research over the next few years.

Meteorological observations at manned stations will be increasingly supplemented by surface observations from automatic stations (see Chapter 9), which will increase the data coverage over the Antarctic ice sheet, and by improved remote-sensing of atmospheric parameters from satellite. Automatic stations, together with other observing systems, are also being used to study medium-scale processes and to help to define the constraints on boundary fluxes in katabatic winds. Similarly, more automatic observations can be made from drifting buoys in the Southern Ocean and from buoys and stations deployed on icebergs and sea ice within the seasonal sea ice zone.

An experiment was run in the Weddell Sea in 1986 by a team of scientists from various SCAR nations, aboard a West German research vessel. The experience gained in the Marginal Ice Zone

Soviet ice drill in use on the Ross Ice Shelf

Experiment (MIZEX) in the Arctic was transferred to the Antarctic with a comprehensive study of ice movement, thickness, energy fluxes and atmospheric driving forces. The behaviour of air masses along the eastern shore of the Weddell Sea was studied from the ship and from land stations within a few hundred kilometres of each other and the ship.

With the increasing database, it will be possible to improve definition of the constraining factors in sophisticated computer models of climate and to validate the models. If the performance of these models can be improved, the range of Antarctic climate problems that can be tackled by numerical experimentation is widened.

Observations of a range of trace gases and particulate materials are expected to continue and to expand. Such observations will provide indicators of potential changes in climate and enable atmospheric constituents to be related to impurities in deposited snow.

SATELLITE MONITORING AND DRIFTING BUOYS

It is vital that the satellite monitoring of the extent and concentration of sea ice continues, and that systematic and consistent analysis of the data is undertaken. At the same time, the development of new satellite systems offers the possibility of measuring other variables from space. Multichannel, higher resolution, passive microwave radiometers should make it easier to distinguish between the various types of ice from space, and active microwave radar systems can provide high-resolution details of ice morphology, structural changes and drift.

The observation of ice characteristics and properties within the seasonal ice zone is expected to be a major research area, not only for studies related to climate but also because of the importance of the sea ice to biological and oceanic processes. Several nations are planning winter research cruises within the ice zone. Automatic drifting buoys will define the broad features of Antarctic ice drift and provide data on meteorological and oceanic variables. It should be possible to integrate some simple observations of ice characteristics during resupply voyages and flights to research stations.

DEEP DRILLED ICE CORES

Future research on ice cores is expected to continue along the present lines. Improvements in the technology of drilling now make it possible to recover very deep ice cores. Soviet scientists, as we have already seen (p. 85), have recovered a core from 2084 m at Vostok. Plans are now being formulated to drill completely through the ice sheet to the underlying rocks in up to 4000 m of ice.

Cores of shallow and intermediate depths will provide data to create a regional time series of climatic variables for the past century and the

Sampling a core of ice near Syowa Station in winter

most recent millennia. Ice cores to bedrock in West Antarctica would provide evidence of the likely response of that marine ice sheet to possible climatic warming induced by carbon dioxide (Fig. 10.5).

New techniques and instruments make it possible to analyse ice cores in more detail and with greater accuracy than before, and the development of combined accelerator/mass spectrometer systems makes it possible to measure cosmogenic radio-isotopes (such as carbon-14) for much more

Antarctic Peninsula (GAP) and the Filchner/ Ronne Ice Shelf Project. Studies are also in hand on the dynamics of ice sliding in large outlet glaciers and West Antarctic ice streams, and also on melt/freezing processes below major ice shelves in the Ross and Weddell Seas.

Definition and monitoring of the ice front for any changes can be undertaken from imagery from satellites such as LANDSAT-5 and SPOT, and further high-detail mapping of the elevation of the ice sheet would be possible with data from

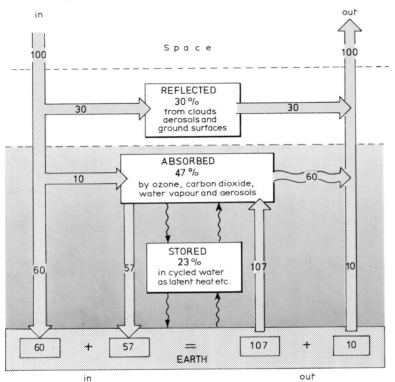

Fig 10.5 Radiation exchanges and balances. Carbon dioxide and water vapour in the atmosphere are virtually transparent to incoming solar radiation. A build up of carbon dioxide or water vapour in the atmosphere would trap increasing quantities of outgoing heat. This would lead to a rise of surface temperature and thus to global warming —the greenhouse effect

accurate dating. Correlation of the ice-core record with meteorological data for the past 30 years, and joint studies of atmospheric aerosols and isotopic composition and impurities in new snow, will help to 'calibrate' the ice-core record.

MULTINATIONAL COLLABORATION

A number of large-scale, multinational collaborative programmes aimed at determining the present characteristics and dynamics of the ice sheet will continue. These include the International Antarctic Glaciological Project (IAGP), studies on the Ross Ice Shelf, the Glaciology of the

radar altimeters planned for satellites by the American and European space agencies. Mapping by radar altimeters of the smooth central regions of ice sheets at intervals of 10 years will show real surface changes and assist in evaluating the long-term mass budget.

EFFECTS OF CHANGING CLIMATE

Climate is a primary factor in all environments. Research indicates that climatic change affecting the Antarctic would not only have a local effect but also that interactions between the atmosphere, ocean and cryosphere could lead to feedback affecting the whole Earth. Simulations with computer models suggest that the polar regions are uniquely vulnerable to climatic change of the kind projected to occur as a result of increases in atmospheric temperature.

Man's activities are responsible for enhanced levels of carbon dioxide and other gases that affect the Earth's radiation balance and could result in increased atmospheric temperature and significant decreases in Antarctic sea ice. The resultant feedback consequences, could occur in a matter of decades and, on the longer time-scale, the stability of the main Antarctic ice sheet could be affected.

The continued monitoring of trace gases and particulate material in the Antarctic atmosphere, the monitoring of the extent of Antarctic sea ice as a sensitive climatic indicator, and the study of the stability of marine ice sheet are all important to the protection of the Antarctic and global environment.

11

The upper atmosphere from the Antarctic

At distances beyond about 64 000 km, or 10 times the Earth's radius, a supersonic stream of charged particles, or 'plasma', continuously buffets the outer environment of the Earth. This is the solar wind, consisting of electrons and ions, mainly hydrogen ions (protons). It has a typical density of a few million electrons per m³, and it flows from the Sun at several hundred kilometres per second.

The Earth's magnetic field extends into space, creating the region called the magnetosphere (Fig. 11.1). The magnetosphere protects the Earth, from bombardment by charged particles, except in the polar regions, and deflects the flow of the solar wind around the planet. The outer boundary of the magnetosphere, termed the magnetopause, forms where pressure of the solar wind balances the pressure due to the Earth's magnetic field. On the nightside of the planet, the flow of the solar wind drags the geomagnetic field lines out into a comet-like tail, which extends for some million kilometres downstream.

The plasma in the magnetosphere comes from both the solar wind and the ionosphere. The ionospheric plasma, which is produced mainly at altitudes between 70 and 250 km by ultraviolet and X-radiation from the Sun, is of low thermal energy

Aurora viewed from the Japanese station Syowa

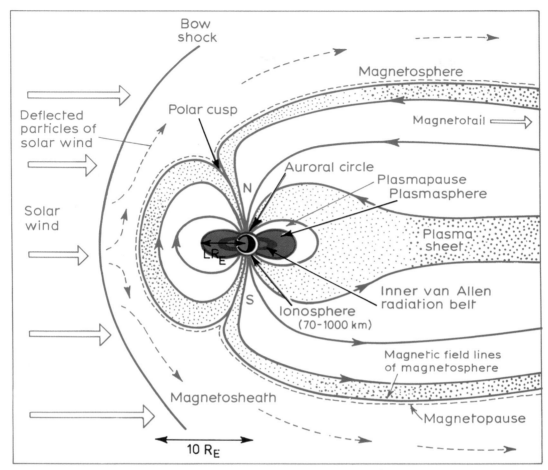

Fig 11.1 The magnetosphere acts as an obstacle in the supersonic flow of the solar wind. Upstream of the magnetosphere lies the bow shock, where the solar wind is slowed down. The diagram also shows regions of particular interest and, further, illustrates how a field line is specified by its *L* value. This is effectively the distance, measured in Earth radii, R_E, from the centre of the Earth to where this field line crosses the magnetic equatorial plane

(around 1 electron-volt, or 2×10^{-19} joules), and is the dominant source of the region known as the plasmasphere, shown in Fig. 11.1. The plasmapause, the outer boundary of the plasmasphere, usually lies at a distance in the equatorial plane of about four times the Earth's radius away from the Earth's centre. Here the geomagnetic field is similar to that of a dipole, rather as if a bar magnet were buried near the centre of the Earth. Some plasma of higher energy from the solar wind penetrates the magnetosphere, and reaches the upper atmosphere above the poles. In particular, it does this by way of the polar cusps, which separate the closed geomagnetic field lines on the dayside from those open magnetic field lines that stretch back into the tail and are joined to interplanetary magnetic field lines. Plasma from the solar wind is accelerated by various mechanisms and is stored in the plasma sheet in the magnetospheric tail. Nearer to Earth, the plasma is also stored in the van Allen radiation belts, where the charged particles bounce back and forth along dipolar geomagnetic field lines. Such charged particles release their energy when magnetic disturbances, some induced by solar activity, cause them to drop out, or 'precipitate', into the atmosphere at high latitudes. The interaction of the solar wind with the Earth's magnetic field creates a wide range of effects—aurorae, geomagnetic storms, disruption in shortwave-radio communications, power surges in long electricity transmission lines—which are collectively called solar-terrestrial phenomena.

ANTARCTICA AND GEOSPACE

'Geospace' is the term now used to describe the region where the solar wind interacts with the Earth's magnetic field and atmosphere.

Electrons moving along the outer edge of the plasma sheet are subject to complex processes which accelerate the particles to energies of a few thousand electronvolts (keV), or about 10^{-15} joules. These electrons bombard the upper atmosphere of the Earth, causing the spectacular aurora australis, in the Southern Hemisphere, and the aurora borealis in the North, at heights around 110 km and at geomagnetic latitudes around 67° (Fig. 11.2). Geospace can, in fact, be viewed as a gigantic television tube. The centre of the magnetospheric tail corresponds to the electron gun, the polar upper atmosphere to the screen of the tube, and the aurora to the image on the screen. The auroral displays are created by electrons coming in and colliding with atoms and molecules in the upper atmosphere. The molecules then emit photons. The intensity of the display depends on the intensity of the solar wind and on both the magnitude and direction of the magnetic field that it pulls out from the Sun. In this way, the effects of 'distant' magnetospheric processes are revealed 'nearby'—in the upper atmosphere of the northern and southern polar regions. Consequently, the

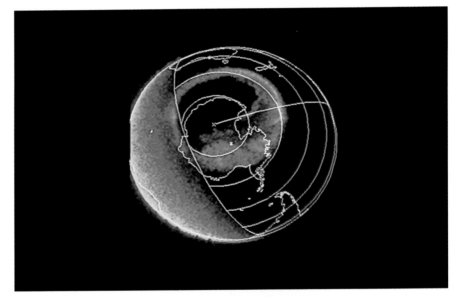

Fig 11.2 The auroral oval viewed from the Dynamics Explorer-1 satellite. The auroral oval is a continuous, almost circular, region centred on the geomagnetic pole

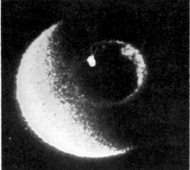

Sequence of events in an auroral storm taken 24 minutes apart from NASA's Dynamics Explorer spacecraft on 15 September 1981. Photo A is representation of a quiet period; photo B is peak storm; and C is the recovery from the storm

polar regions provide splendid bases for studying magnetospheric phenomena, in conjunction with observations made from rockets and spacecraft orbiting the Earth.

The advantage of Antarctica over the northern polar regions for these studies is that it is not an ice-bound ocean like the Arctic. Scientists have set up permanent observatories on the Antarctic continent, both staffed and automated (Fig. 11.3), to enable them to investigate phenomena occurring at great distances from Earth, for example at the plasmapause, above the auroral regions, in the polar cusp and in the plasma sheet (Fig. 11.1).

and Siple—lie at the foot of the geomagnetic field lines along the plasmapause when the geomagnetic activity is 'moderate'. The plasmapause is an important surface aligned with the magnetic field. It marks the outer boundary of the plasmasphere—the relatively cool (low energy) plasma of ionospheric origin that rotates with the Earth. As we have already seen, charged particles from the solar wind are largely swept along the magnetopause and slowly flow inwards to add to the plasma sheet far behind the Earth. As a result, hot plasma builds up in the comet-like tail of the magnetosphere and becomes driven in the equa-

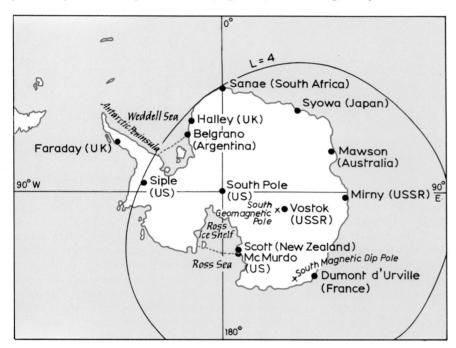

Fig 11.3 The key stations in Antarctica for upper atmosphere and ionospheric physics, and the loci of the feet of the geomagnetic field lines at $L=4$

Note Since this map was drawn, the South Magnetic Dip Pole has moved offshore. By January 1986 it was at 68° South and 140°01′ East

The Japanese Syowa Station on Antarctica, for example, is a prime location for investigating auroral displays. Events on the Sun produce gusts in the solar wind and, when its speed is higher than usual, the electric currents that flow in the ionosphere, and which are responsible for geomagnetic activity, become enhanced. In such circumstances, the magnetosphere is compressed and the auroral regions are nudged towards the lower magnetic latitudes. During such 'active' conditions, the aurora can be overhead at stations such as Halley and Siple which are at a geomagnetic latitude of 61°.

Three stations on Antarctica—Halley, Sanae,

torial plane back towards the day-side magnetosphere at speeds of a few kilometres per second. The build-up and flow of plasma in the tail is as if there were two vast cells of plasma circulating through the magnetosphere, one in the North and the other in the South.

As Fig. 11.4 illustrates, charged particles precipitate into the auroral zone, on the geomagnetic pole side of the plasmapause. Auroral charged particles can further produce thin layers with greatly enhanced concentrations of electrons. These patches of so-called sporadic E ionisation (E_s) give rise to anomalous reception of high-frequency radio waves from ground transmitters.

Fig 11.4 Interesting thermospheric and ionospheric phenomena occur near the mid-latitude trough, that is close to the geomagnetic flux tube along the plasmapause. The 'invariant co-ordinates' are the co-ordinates of the magnetosphere, derived from L (see Fig 11.1), when projected onto the surface of the Earth

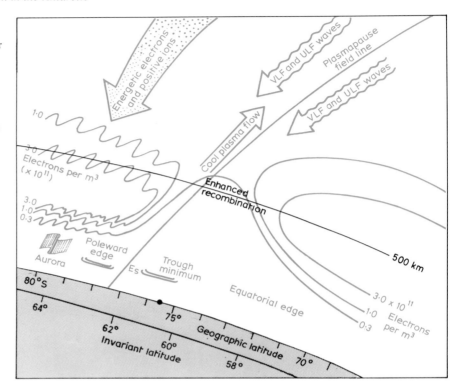

The auroral particles also heat the uppermost atmosphere, causing a local wind to blow there.

Studies of geospace provide practical benefits not only for radio communications but also for geophysical (magnetic) exploration at high latitudes, particularly during the magnetically disturbed conditions that exist during and after auroral activity. The electrically neutral upper atmosphere (the thermosphere) at a height from 110 km up to about 300 km, also responds to auroral activity and becomes markedly warmer. Consequential increases in density affect the orbits of polar-orbiting satellites at low altitudes.

Towards the foot of magnetic field lines through the plasmapause, the density of electrons in the plasma about 300 km up in the ionosphere (the F2 region) falls during night-time to a minimum; this is known as the mid-latitude electron-density trough. The arch-shaped contours of the electron density in Fig. 11.4 show the presence of this trough. The trough is believed to form because ions and electrons in the plasma are recombining faster than usual, since the temperature is higher, and because the plasma is moving out of the region.

Instruments called ionosondes have been used to study the ionosphere at a number of locations in the Antarctic since the IGY. These radar devices transmit pulses of radio waves of varying frequencies up to the ionosphere and detect any echoes produced as the radio waves are reflected by the ionized layers of the atmosphere. The ordinary wave pulses are reflected at levels in the ionosphere where the 'plasma frequency' equals the radio frequency. Thus, if a known radio frequency is reflected from an ionospheric level, the electron concentration at that level can be calculated. From the travel time of the pulse, the height of the reflecting level can be calculated.

Since 1981, computer-controlled ionosondes called the Advanced Ionospheric Sounders (AIS) have been operated at Siple and Halley stations (Fig. 11.5). The ionospheric echoes are automatically digitally processed by the AIS computer. The AIS at Siple and Halley have been used to study the evolution and dynamics of the mid-latitude electron density trough (Fig. 11.6, upper right). Measurements have also been made in conjunction with the US Dynamics Explorer satellites, which provide *in situ* observations of not

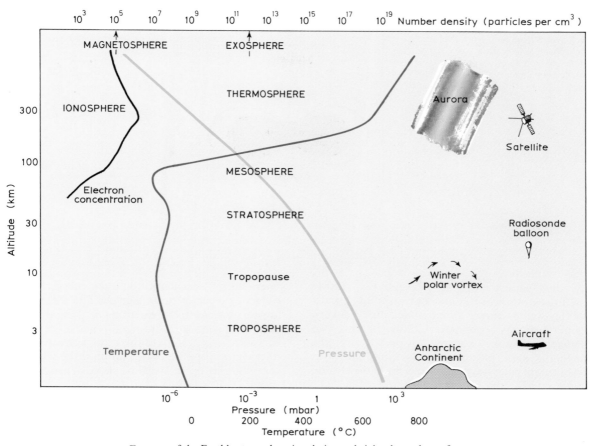

Features of the Earth's atmosphere in relation to height above the surface

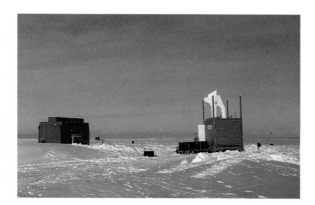

Fig 11.5 The Advanced Ionospheric Sounder, at Halley Station, is in the hut on the left and the diesel electric generator is in the hut on the right

only the ionospheric but also magnetospheric plasmas.

The plasmapause is marked by a sudden drop in the electron density by a factor of 10 or more in the equatorial plane (Fig. 11.6, lower right). It usually occurs within a narrow region only a fraction of an Earth radius thick. The position of the plasmapause can be delineated by natural radio waves of very-low-frequency (VLF), particularly by the so-called 'whistlers'. Scientists working in the Antarctic can receive whistlers, radio signals at audible frequencies that are generated by lightning flashes in the Northern Hemisphere at the other end of magnetic field lines in Antarctica. Some radio energy propagates through the ionosphere and is guided by 'ducts' of somewhat enhanced electron density. These ducts, which are aligned along geomagnetic field lines,

Fig 11.6 Left, Meridian view of the plasmasphere (blue). Top right, The L-shell profile through the F region of the ionosphere, and, bottom, right, the profile of the electron density with L in the equatorial plane

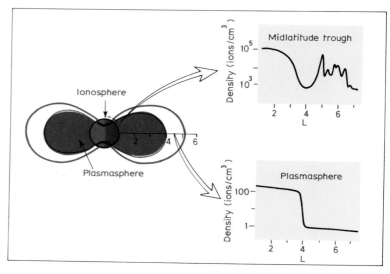

are thus said to act as 'waveguides'—rather like fibre-optic wave-guides but on a huge scale. As the signals travel along a field line between geomagnetically conjugate points in the two hemispheres they become dispersed by the magnetospheric plasma; that is, components at lower frequencies normally travel more slowly and arrive later at the far end of the field line. The signals are thus heard as tones of falling frequency, hence the term whistlers.

Whistlers yield information on the regions close to the equatorial plane through which they have travelled and where most of the dispersion occurs. The signals arriving at Siple, Belgrano, Halley, or Sanae stations have passed in their travels from the Northern Hemisphere out to regions around 4.2 Earth radii from the centre of the Earth. Studies of whistlers in Antarctica provide interesting information on the position of the plasmapause, on 'ripples' on the plasmapause and on longitudinal structure in the plasmapause.

Whistlers heading down towards the Earth's surface can be reflected by the ionosphere or by the ground, forming 'multi-hop' whistlers. Observations of the same whistlers at Halley, Siple, and Palmer stations have shown that whistlers can emerge from one electron density duct and be refracted into another some 300 km away. Such coupling between ducts can occur from inside the plasmapause to outside it, where the whistler can trigger tones of rising frequency, termed 'chorus'. Chorus often appears to arise spontaneously, beyond the plasmapause, and is believed to be

generated by a resonance between energetic electrons and whistler radio signals. The interaction transfers energy from electrons in the van Allen belts to the whistler radio signals, and so amplifies the waves. Interactions of this kind between coherent waves and particles can amplify the whistlers themselves by up to 30 dB per hop, depending on the purity of the signals.

Atmospheric scientists have recently discovered that electromagnetic radiation due to power lines in the electrical grid system of industrialized areas can modify processes in the magnetosphere. Such effects represent a new form of pollution. Several stations in Antarctica at locations that hitherto were electromagnetically 'quiet' are now found to be receiving the signals that arise from such pollution in the industrialized parts of the Northern Hemisphere.

Another topic of recent research co-ordinated by SCAR is the Trimpi event, named after its discoverer. This is the observation (Fig. 11.7), following a whistler, of changes in the amplitude and/or the phase of signals from a VLF transmitter which are propagating in the waveguide between the Earth and the ionosphere. A Trimpi event is believed to be due to an interaction which transfers energy from the electrons to the whistler. This happens in such a way as to reduce the angle between the direction of motion of the electrons and the geomagnetic field. The electrons can thus precipitate out of the radiation belts into the upper atmosphere where they produce additional ionisation in the ionosphere about 80 km up (within

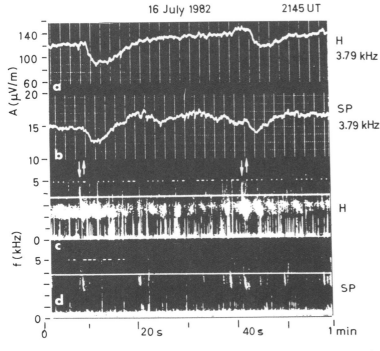

16 July 1982   2145 UT

Fig 11.7 Top panels show the amplitude of the 3.79 kHz signal from the transmitter at Siple and recorded at Halley (H) and South Pole (SP). At times of whistlers, ↓, whose spectrograms at Halley and South Pole are shown in the lower panel, the Trimpi events can be seen at ↑

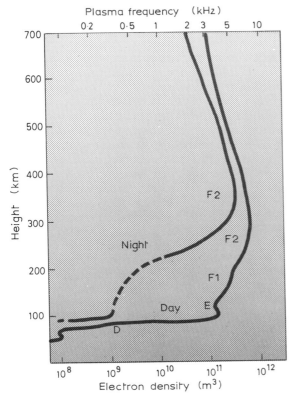

Fig 11.8 Typical ionosphere electron-density profiles by day and night

the D region in Fig. 11.8). Recent observations indicate that a signal from a tunable VLF transmitter at Siple (Fig. 11.9) can detect Trimpi events. Moreover, this transmitter can stimulate emissions that in some respects resemble chorus. Such experiments have led to considerable theoretical and computational studies of the plasma physics of the interactions between waves and electrons. These studies have a far wider application than just this single experiment, for example in other natural cosmic plasmas and also in laboratory plasma devices.

At altitudes above about 85 km, the temperature in the atmosphere begins to rise, reaching even 1500° C in the thermosphere. Here there are winds of un-ionized, electrically neutral atoms and molecules. These neutral winds are investigated by various techniques from Antarctic stations. For example, they can be detected by so-called 'drift radar' operating at a frequency of 2 MHz at Mawson station. The neutral winds are also measured from Mawson and Halley using instruments called Fabry–Perot interferometers, which observe red light (wavelength 630 nm) and green

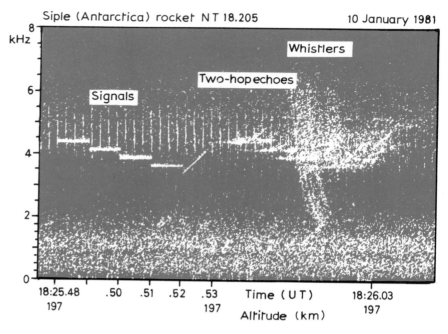

Fig 11.9 Dynamic spectrum of VLF radio signals received by a loop antenna mounted on a rocket launched from Siple Station

light (wavelength 755 nm) emitted by excited, energized, oxygen atoms at an altitude of 95 km. These emissions may also provide estimates of temperature for these levels. Ions blown by the winds move easily along geomagnetic field lines, and observations of the maximum density of the plasma in the F layer are generally consistent with the measured neutral winds.

Another experiment conducted at many stations around Antarctica and on sub-Antarctic islands uses a riometer (*r*elative *i*onospheric *o*pacity *meter*) to measure the strength of incoming radio noise from the cosmos at a frequency of 30 MHz. The intensity of the received noise varies systematically as the Earth rotates, but there can be additional transient decreases in the strength of the signal. These may be due to extra absorption in the ionosphere (mainly in the D region) caused by additional ionization produced by energetic charged particles that are released from the van Allen belts. Differences in the timing of such transient events provide information about how the charged particles are bouncing back and forth across the equator and drifting in longitude.

In the Southern Hemisphere, the geomagnetic pole is twice as far from the geographic pole as it is in the North. At a particular geographic latitude in Antarctica there is, at different longitudes, a large variation of geomagnetic latitude. Because many Antarctic stations are situated at approximately the same geographic latitude around the coast, this feature, taken together with the international co-operation that characterizes Antarctic science, provides a unique opportunity for studies of upper atmospheric physics at a range of magnetic latitudes. In particular, the sector of Antarctica from 0° to 90° West (see Fig. 11.3) forms a natural laboratory for studies of ionospheric phenomena associated with the plasmapause. It is only in this sector of Antarctica that phenomena related to the plasmapause can be studied at such high geographic latitudes. The geometry of the geomagnetic field in this region makes the F region of the ionosphere particularly sensitive to movements due to electric fields or neutral winds, or both. This explains the very different behaviour of the F region electron densities that are evident in Fig. 11.10. The stations Faraday and Dumont d'Urville are at a similar geographic latitude but are on diametrically opposite sides of the South Pole (see Fig. 11.3). The dip angles of the geomagnetic field differ by more than 30° at the two locations. From Faraday, the magnetic field lines go into the plasmasphere; however, from Dumont d'Urville, near the dip magnetic pole where the geomagnetic field lines are vertical, they run out into the magnetospheric tail.

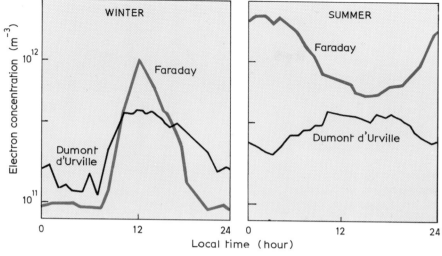

WINTER

Faraday

Dumont
d'Urville

SUMMER

Faraday

Dumont d'Urville

Electron concentration (m⁻³)

Local time (hour)

Fig 11.10 Markedly different diurnal variations of the electron density in the ionosphere at the peak of the F2 region discerned at Faraday and Dumont d'Urville

The long winter nights within the Antarctic Circle, when there is no direct ionization by the Sun's radiation, provide ample opportunity to study features of the ionosphere that are tied to the geomagnetic field. In the summer months, however, the upper atmosphere receives a more or less constant amount of solar energy which produces ionization at a steady rate. Thus, within the Antarctic Circle, the *annual* variation in the solar radiation, and in the electron density in the ionosphere that such radiation creates, is rather like the *daily* variation more familiar away from the polar regions.

Scientists at several Antarctic stations have measured the geomagnetic field for many years. These records provide annual mean values that contribute to the basis of world-wide and up-to-date charts used for navigation. The geomagnetic values change, indicating movements of the magnetic poles and variations in the strength of the field caused by changes occurring deep within the Earth. On shorter time-scales, variations occur due to changing ionospheric currents and changing patterns in the precipitation of charged particles. At even shorter periods of around a minute, pulsations in the geomagnetic field give

Soviet scientists establish geomagnetic observatories in Antarctica

information on the plasma density in the magnetosphere. Moreover, recent discoveries of pulsations with periods of several seconds, observed at Davis, Casey, Mirnyy, and Amundsen–Scott stations have provided new insight into the direct penetration of star wind energy into the high latitude ionosphere.

The Antarctic Peninsula provides unique opportunities to take further advantage of the large differences between geomagnetic and geographic co-ordinates at magnetic mid-latitudes. There is a similar large difference between these co-ordinates at the northern end of the field lines from the Antarctic Peninsula region; in North America the geomagnetic latitude is greater than the geographic latitude whereas it is vice versa in the South. This makes studies at the two ends of the field lines particularly important at these longitudes. There are further possibilities for future scientific collaboration between Argentina, Chile, the United Kingdom and the United States in the Antarctic Peninsula, and with Canada and the United States in the Arctic.

Exciting possibilities for future SCAR projects in the Antarctic include an array of automated geophysical observatories, together with one or

A Nike-Tomahawk rocket is launched from Siple Station in January 1981

Launching a scientific balloon for upper atmosphere research at Syowa Station

more radars to view an extensive area including the auroral oval and the polar cap in a region that is at the opposite end of geomagnetic field lines from the much-studied ionosphere over Greenland. Rocket and other radar experiments on the dynamic atmosphere over Antarctica will undoubtedly lead to important new results of global significance, as will experiments carried aloft aboard polar-orbiting satellites, such as the proposed Polar Platform component of the US Space Station.

ICSU scientists recently proposed plans for a concentrated and co-ordinated study of plasma interactions in the Sun–Earth system; this knowledge may be extrapolated to the other planets and to the Universe beyond. The study, to be known as the International Solar-Terrestrial Physics programme, is intended to begin in the early 1990s. A major component of the programme is to be the Global Geospace Study, proposed by the US National Aeronautical and Space Administration. The European Space Agency and the Institute of Space and Astronautical Science (Japan) also plan to launch scientific satellites for this international programme. Besides its focus on the workings of the Sun and the origin of the aurorae, the programme has considerable potential for practical applications involving the Earth's environment, including the increasingly routine operation of near-Earth spacecraft.

12

Marine life and the living edge of Antarctica

The SCAR Working Group on Biology, in addition to promoting and co-ordinating biological research in Antarctica and in the Southern Ocean, has played an active role in helping to draft the conservation measures embodied in the Agreed Measures for the Conservation of Antarctic Flora and Fauna, under the Antarctic Treaty, and in the intergovernmental conventions on the Conservation of Seals and Marine Living Resources. Its Conservation Subcommittee has been most active in advising the working group how it might respond to requests for advice on conservation addressed to SCAR by the Antarctic Treaty. Two other Groups of Specialists have reported on environmental consequences of possible mineral exploration and exploitation. This progress has been achieved through the close collaboration between all those studying the biology of the Antarctic, and this collaboration is assured, at remarkably low cost, through SCAR.

The Group of Specialists on Southern Ocean Ecosystems and their Living Resources was established as a result of an initiative taken by the Working Group on Biology, and there has been close liaison between the two groups through cross-membership. The Subcommittee on Bird Biology has acted as the co-ordinator of the International Survey of Antarctic Seabirds (ISAS) which was an integral component of the BIOMASS Programme (see below). The Group of Specialists on Seals has, among much else, the task of monitoring seal populations.

THE ANTARCTIC ENVIRONMENT

The Antarctic region presents a range of what seem at first sight improbable habitats for life, on its frozen shores, in its cold seas, on its ice shelves and on the sea floor beneath them, and in its extensive zones of pack ice.

The extremes of geographic setting and climate create unusual conditions. For example, during the southern summer the intensity of sunlight is

Rookery of Adélie penguins on Rumpa Island, South of Syowa Station

high, yet air temperatures remain near to zero even though the region receives more solar radiation in a day than reaches the Equator in an equivalent period. The main reason is that much of the solar radiation is reflected back into space by the ice sheets.

Other extremes include the fact that, in winter, night in the more southerly coastal regions of Antarctica may be two or three months long: beyond latitude 80° South, on the pack ice or inland, night lasts four to five months—and six months at the South Pole. Antarctica offers good opportunities to study how life has adapted and has evolved strategies to cope with extreme environments.

The Antarctic terrestrial environment and its associated plants and animals constitute the least changed natural system to be found anywhere in the world. The Antarctic Treaty system recognized early the importance of Antarctic life in its Agreed Measures for the Conservation of Antarctic Flora and Fauna. Such recognition was extended to the high seas first by the Convention for the Conservation of Antarctic Seals (CCAS), in 1972, and more recently, by the comprehensive Convention for the Conservation of Antarctic Marine Living Resources (CCAMLR) of 1980. The Working Group on Biology is currently looking at a new programme of research focusing especially on terrestrial biology—the Biological Investigations of Terrestrial Antarctic Systems (BIOTAS).

The living resources convention has conservation objectives that are very new in concept. Specifically, the convention aims to control exploitation of finfish and the shrimp-like krill, *Euphausia superba*, the key organism in the food web and staple food of many fish, penguins, seals, and whales. However, the convention is also concerned with the species that depend on the krill, and with the maintenance of their ecological relationships, for one of the big differences between the Antarctic and elsewhere is the presence of very abundant and conspicuous large secondary consumers: the seals, whales, and birds.

THE ANTARCTIC MARINE ECOSYSTEM

The Southern Ocean is an unusual and highly specialized habitat, showing great seasonal fluctuations in the ratio of open to ice-covered sea, with minute algae (often associated with ice), krill and other invertebrates, fish, birds, seals, and whales forming relatively short food chains. It is a large semi-enclosed unit, probably the largest marine ecosystem on the globe. The northern boundary is fairly well defined by the Antarctic Convergence in the upper 500–1000m. Whales and birds cross the convergence regularly and are responsible for the 'export' of a certain amount of organic matter. In the deeper layers there is a meridional transport of nutrients and zooplankton by the relatively warm Deep Water moving southwards and by the Antarctic Bottom Water moving northwards (see Chapter 8). At its continental boundary, the Southern Ocean, unlike other oceans, is not influenced by terrestrial biota

and river input. The peculiarity and adaptation of many of the species of the Southern Ocean are indications of a long evolutionary process in isolation. It is considered that the circulation patterns and water masses were established in their present form at least 12–15 million years ago. As a consequence, the major taxonomic components are circumpolar in their distribution.

Because of the upwelling in the region of the Antarctic Divergence along the boundary between the East Wind and West Wind Drifts (see Fig. 8.4) and of other vertical processes, nutrients generally are not limiting for the growth of phytoplankton. Primary production appears to be limited mainly by light conditions. Turbulence induced in the water column by wind and freezing processes reduces the light to which algal cells are exposed, and may prevent the cells from making full use of the nutrient supply. Furthermore, about half the total area of the Southern Ocean is covered with pack ice at the end of the winter; in the austral summer this ice shrinks to a 10 per cent coverage. Important effects of the ice cover are the shading of the water beneath, the surface layer of melt-water in the spring and summer, and the liberation of large amounts of ice-algae at the break-up and melting of the pack ice. The ice-algae have been estimated to provide between 15 and 20 per cent of the total primary production. Of this about 40 per cent probably gradually enters the water column through the ice-algae/ice-invertebrates, krill, larval fish/fish food chain (see below). Year-to-year variations in the extent and timing of the ice-cover have a very large effect on the otherwise stable conditions of the Southern Ocean.

Studies of the zooplankton of the Southern Ocean, especially krill (*Euphausia superba*), and of their consumers, fish, squid, birds, seals, and whales are important both for their intrinsic scientific interest, and for providing a basis for advising on the rational management of resource exploitation and conservation of the Antarctic marine ecosystem. The international programme of Biological Investigations of Marine Antarctic Systems and Stocks (BIOMASS) is especially important in this respect.

SOUTHERN OCEAN FOOD WEBS

As a result of recent research, traditional views of the productivity of the Southern Ocean have changed. High primary production in the shelf region and near the pack ice and the short

The Antarctic shallow-water, bottom living communities are dominated by attached invertebrates and an associated community of browser/scavengers of which echinoderms are well represented in terms of number and biomass. The most evident species in this montage are sea cucumbers, sea urchins, star fish, brittle stars, sea squirts, and red/purple encrusting algae

diatom—krill—whale food chain were first taken as indicators of a very large renewable resource. However, it is now known that primary production in the open parts of the Southern Ocean is relatively variable, low in winter, high in summer, and the stocks of krill are now thought to be considerably less than earlier estimates. Recent research has yielded a more detailed and realistic picture of the distribution and abundance of some of the key elements in the food webs of the Southern Ocean, but there is much that is not known with certainty.

The pelagic Southern Ocean contains three major zones which, although they interact, may for convenience be considered as having distinct food webs. These zones are: the ice-free zone of the West Wind Drift, dominated by herbivorous copepods, salps, and small euphausiids; the seasonal pack-ice zone of the East Wind Drift and adjacent fronts and eddies in which krill is the dominant element; and the permanent sea-ice zone near the Antarctic continent and its ice shelves where zooplankton biomass (total mass) is low and, thus, predators are scarce.

The ice-free zone of the oceanic West Wind Drift is rich in nutrients but relatively poor in primary production and phytoplankton biomass. The zooplankton is similar to that of the northern North Atlantic. Herbivorous copepods, salps, and small euphausiids are dominant. Except for around South Georgia, krill is mostly absent. Squid, myctophids, and some juveniles of benthic fish species are present, but play a minor role in the West Wind Drift compared with the shoaling fish of the North Atlantic and North Pacific. Flying birds tend to concentrate in the region of the Antarctic Convergence.

The seasonal pack-ice zone is covered by ice in

winter and spring, but it is mainly ice-free in summer and autumn. It occupies most of the East Wind Drift and the large eddies along the Antarctic Divergence, and includes the northern branch of the Weddell Gyre and the waters off the Antarctic Peninsula. Large parts of the zone are comparatively shallow, particularly in the Atlantic sector. Possibly stabilization of the upper layers in the vicinity of the pack ice and seeding by ice-algae make this zone the most productive one on a yearly basis. With the breaking up and retreat of the pack ice in spring and summer, a series of phytoplankton blooms may proceed from North to South. Large amounts of phytoplankton sink to the bottom, out of this system. The system comprises a complex food web with salps, copepods, euphausiids, fish larvae, and chaetognaths. However, krill is the dominant element. Shoals of krill are sufficiently large and dense to provide the food base for baleen whales, crabeater seals, and Adélie penguins, at low-energy cost.

The permanent pack-ice zone extends over the area of the cold near-shore water masses (ice-shelf water) which are partly separated from the East Wind Drift by the Continental Convergence. These water masses are particularly well developed in the shallow parts of the inner Weddell Sea and in the Ross Sea. They are the habitat of a biotic community with a very intense, but brief, production period. Krill is replaced by the small *Euphausia crystallorophias*. Accordingly, krill-eating mammals and birds are scarce. Pelagic fish, particularly *Pleuragramma antarcticum*, seem to be more abundant than elsewhere in the Southern Ocean. Most of the primary production by phytoplankton and ice-algae goes to the suspension feeders, such as sponges and echinoderms. They are the food source of certain crustaceans, cephalopods, and of a large number of fish species which are taken by emperor penguins, and Weddell seals.

BIOLOGICAL SEA-ICE STUDIES

The physical and biological changes that occur during the seasonal advance and retreat of the pack-ice are dramatic. The hypothesis has been advanced that ice-edge zones are major sites of primary production and energy transfer within the Southern Ocean, and that large accumulations of the upper-trophic-level organisms (birds and mammals) occur in these zones.

The sea-ice is the resting place for seals and

Antarctic krill

penguins from where they hunt for krill and fish. Recent studies have demonstrated that the sea-ice is a microcosm on its own. It is not only the habitat of unicellular algae staining the ice yellowish-brown, but also of many different kinds of animals such as foraminifera, rotifers, nematodes, and crustacea. Even small fish live in the lacunae and feed on small organisms attached to the ice floes. Many microorganisms can stand the low temperatures and high salinities of the brine channels and pockets within the porous sea-ice.

KRILL

The standing crop of zooplankton in the Antarctic waters is significantly higher than in tropical and most temperate regions. This is partly due to the high abundance of *Euphausia superba*, the Antarctic krill. Of the consumers of krill, the seals are the most important; next in the consumer order are the birds. At one time, the baleen whales, such as the blue and fin, ate the largest share of the krill, but with the decline in the cetaceans due to commercial whaling they have been consigned to third place. Squid may now be consuming more krill than the whales, but our knowledge of the ecology of these animals is very sparse indeed.

In recent years, considerable advances have been made in understanding the swarming behaviour of krill. Acoustic methods have made it clear that the krill can form huge swarms ('super-swarms') that may be several kilometres across but often only 5 or 10 metres thick. A superswarm may contain several million tonnes of krill. These new finds point to the difficulties of estimating with any accuracy the standing stock of krill— because the very patchy distribution introduces enormous variation in the estimates.

The international BIOMASS programme has been important in assessing the stocks of krill. The data acquisition phase of this 10-year programme (1976–1986) involved several nations and included two peaks of collaborative multiple-ship investigations. The first of these was FIBEX (First International BIOMASS Experiment) in early 1981 in which 13 ships of 10 nations worked together in a co-ordinated programme covering a large area of the Scotia Sea, the Drake Passage, and parts of the Indian and Pacific sectors of the Southern Ocean. This was the largest multiple-ship experiment launched in biological oceanography. It was sponsored by SCAR, SCOR, the Advisory Committee for Marine Resources Research of FAO (ACMRR), and the International Association for Biological Oceanography (IABO). A Group of Specialists on Southern Ocean Ecosystems and their Living Resources was responsible for the development of BIOMASS. The achievements of FIBEX include an acoustic survey of the abundance and distribution of krill in several regions and new information on krill biology, ecology, behaviour, and distribution as well as on phytoplankton, zooplankton, and physical oceanography.

The lessons learned in FIBEX were applied in planning a second experiment, SIBEX, which spanned two southern summer seasons 1983/84 and 1984/85. Seventeen ships of 11 nations were committed to SIBEX.

The data sets of the acoustic surveys, net samples, and oceanographic measurements of all vessels were brought together and jointly evaluated at several international workshops. A BIOMASS Data Centre at the British Antarctic Survey in Cambridge serves as a depository and a working place for further analysis of the data.

What is not yet known with any accuracy is the annual production of krill. Growth has been estimated by several independent methods at amounts that vary. In the surface waters in summer the juvenile krill grows in length from about 35 mm to 45 mm in a matter of a few months while feeding on a range of marine phytoplankton-

and micro-organisms. Growth rate in subsequent years and total life span are less well known.

Much work has been done on the krill's feeding mechanism. Around its mouth is a very complicated feeding basket of fine bristles which forms a cage with which it actively filters the water. It takes in a lot of very small particles, nanoplankton, and even bacteria.

One of the mysteries is what happens in the winter when most of the krill range is covered by pack ice. The region receives little sunlight so there is little phytoplankton. The krill probably feed on detritus in the winter, but they are sometimes carnivorous and even cannibalistic, feeding on small individuals of their own species.

In its third or fourth year of life, krill starts reproduction. The eggs may sink to a depth of 2000 metres or more before the minute larvae hatch and have to struggle upward to find food in the uppermost layers of the sea.

FISH

In recent years, increasing attention has been paid to the fish stocks of the Southern Ocean. Only about a hundred species of fish have been recorded south of the Antarctic Convergence, and of these the dominant group is the Nototheniiformes, comprising five families that make up nearly three-quarters of all coastal fish species. In contrast to other oceans of the world, the Southern Ocean does not appear to contain dense stocks of fishes living near the surface. For many species, especially those subject to exploitation, data have been collected on life cycles, length and age at sexual maturity, natural mortality, age, growth rates, feeding, and energetics. Many of the species lack red blood and are rather sluggish. Slow growth and low fecundity make the Antarctic fish stocks vulnerable to overfishing.

Commercial fisheries are located mainly in the Indian and Atlantic Ocean sectors. So far, 19 species have been recorded in the statistics of the FAO. Most species belong to the subantarctic and Antarctic families Nototheniidae ('Antarctic cod') and Channichthyidae ('ice fish'). Catch statistics have shown that exploitation has followed a similar pattern in both the Atlantic and Indian Ocean sectors. Catches rose respectively to over 400 000 and 200 000 tonnes within one or two seasons, then declined rapidly after two seasons to less than 5 per cent of the peak years. The BIOMASS Working Party on Fish Biology concluded in a recent review of the biology and status of exploited Antarctic fish stocks, that, at the present stage, there are significant deficiencies in the information necessary for a scientific assessment of the stocks and for instituting fisheries management of such stocks. Nevertheless, conservation measures were introduced to certain fisheries around the islands of Kerguelen and South Georgia as a response to the dramatic decline of the local fish stocks.

BIRDS

Research on birds has focused increasingly on their role as predators in the marine ecosystem.

Chinstrap penguin and chick

Such knowledge is now as good, if not better, for the Southern Ocean as for any other marine system. Considering the logistic problems involved, this is no mean achievement.

Many of the projects depend on multinational collaboration, both within and beyond the ambit of **BIOMASS**. This will be a continuing requirement for the future.

Forty species of breeding birds inhabit the Antarctic, but 90 per cent of their total mass and 90 per cent of the food they eat is accounted for by penguins. Of the penguins, the Adélie is the most abundant and feeds on krill. The other major groups of birds are the albatrosses and the petrels. Antarctic biologists have attached time/depth recorders to some of these birds and found that the krill feeders tend to make quite shallow dives to maybe 20 m. Fish and cephalopod feeders make deeper dives.

There are signs around the Antarctic Peninsula of increases in bird populations. It is probable that these increases are related to the decline in whale stocks, which has left the birds facing less competition for food. With the chinstrap penguin, for example, in the South Orkneys, the annual rate of increase since the late 1940s has been about 6 per cent and with the gentoo a rather slow rate at 2 to 5 per cent a year. The king penguin has also increased as have some of the albatrosses, but for reasons that cannot yet be explained, the wandering albatross has decreased. Some petrels showed great increases in the whaling days because of all the offal produced by the whaling operations. Cape pigeons at South Georgia are

Adélie penguins

Emperor penguin and chick

certainly much less abundant now than they were.

Seabird research involving extensive studies on land and the use of various new techniques that have been developed for recording activity at sea is yielding exciting results on range, diet, food consumption, feeding behaviour and bio-energetics. Noteworthy international collaboration includes a pan-Antarctic survey, initiated by SCAR, of selected species of seabirds to locate their colonies and estimate their numbers. A continuing monitoring programme, concentrating primarily on penguins, since they form by far the greatest biomass of the bird populations, will provide information on changes in seabird numbers. The results should also provide an insight into what is happening in the marine ecosystem, using bird numbers as an indicator of change in their prey species.

SEALS

Studies on seals are advancing through the use of newly developed telemetry techniques. These techniques make it possible, by tracking a radio signal from a transmitting device attached to or implanted within the seals, to follow the activities of the animals away from their breeding beaches, and to study their diet and bio-energetics. Assessments of seal stocks and studies of population dynamics of seals also continue on an international basis in different geographical regions. SCAR's Group of Specialists on Seals reviews work on Antarctic seals and the status of the stocks. The group also determines priorities for future research.

Arrival of a column of emperor penguins (about 3000) on the Archipelago of Pointe Géologie

Wandering albatrosses in courtship display

Six types of seal inhabit the seas and shores south of the Antarctic Convergence—two breed on the land, two on the ice and two on the fringe of the ice. The two land-breeders, the fur and elephant seals, have traditionally been of special interest to Man. In fact, the Antarctic fur seal is one of the great success stories in conservation. It was hunted in the 19th century for its luxuriant pelt and it was the basis for the early exploration of the southern oceans. The sealers who sailed down the Antarctic Peninsula were probably the first human visitors to the Antarctic. From 1800 to 1820, and the 1870s, were the two main phases of sealing, and the fur seal had become virtually extinct by the end of the 19th century. Fortunately a few remnant stocks were left at various locations including South Georgia and the Heard and McDonald Island groups. Conservation ordinances were introduced in 1910 for South Georgia and the fur seal was protected from then on. However, even in the 1930s an expedition found only 12 pups on Bird Island and concluded that there were then probably fewer than a couple of hundred fur seals. SCAR scientists of a number of nations have been monitoring numbers of the fur seals since 1956 when the first count indicated there were about 15 000 of them. In 1982 the population was estimated to exceed 900 000. The

Female Antarctic fur seal and pup

increase in numbers will probably begin to tail off as they reach the limits of their food resources. The seals mate in harems and give birth on the beaches. They suckle their young for about 110 days. During this time, the females swim out to a radius of probably 250 km and make a series of trips to feed on krill.

Biologists have fitted recording equipment to the female fur seals to measure the frequency and depths of their dives. The data reveal a close correlation between the behaviour of the fur seals and that of the krill. The krill tend to be nearer the surface at night and at greater depths during the day. The fur seals do most of their feeding at night when the krill is near the surface. The fur seal population is still increasing even though at Bird Island the numbers have probably levelled off. Now satellite colonies are expanding quite rapidly on the mainland of South Georgia and less rapidly farther south in the South Orkney and South Shetland Islands.

The elephant seal, heavy with blubber, was the subject of a second wave of exploitation in the 19th century. While the fur seals provided pelts for fur coats, the elephant-seal hunters were after seal oil, and on various islands they hunted the elephant seal excessively. In 1910, the elephant seal became protected in South Georgia, and the subsequent story is one of remarkable success in rational management. The total world population of southern elephant seal is now about 750 000, the main centres being South Georgia and the islands of Kerguelen, Heard and Macquarie.

Four kinds of seal live in and around the pack ice. These are the proper Antarctic seals. The two at the fringe of the Antarctic pack ice are the crabeater and the leopard seals. The crabeater seal is probably now the most abundant large mammal in the world, numbering about 30 million. For all that its name suggests, it actually eats krill. In the spring the crabeater seals form small breeding groups, consisting of a female and pup which are joined temporarily by a male. Throughout the rest of the year they are widely dispersed around the outer edges of the pack ice.

The distribution of the crabeater overlaps with that of the leopard seal which is much less abundant (about 0.25–0.5 million). The leopard seal is a very catholic feeder—mainly taking krill but also feeding on fish, squid, birds, and other seals. Eighty-five per cent of adult crabeaters have scars that are now known to result from wounds inflicted by the leopard seal.

The least known of all the Antarctic seals is the Ross, which is roughly as numerous as the leopard seal but goes farther into the pack ice than either the leopard or crabeater. The Ross feeds at depth-mainly on squid.

The population of Weddell seals is around 750 000. It is the most southerly mammal, apart from Man, and will spend the winter darkness under the Antarctic 'fast' ice (permanent ice). It gnaws holes in the ice through which it can breathe. The females form colonies on the ice while the males maintain territories in the water. Mating occurs in the water. The Weddell feeds

Crabeater seal on sea ice

Crabeater seals swimming under sea ice

mainly on fish and squid but it also takes invertebrates that live on or close to the sea-bed.

The Antarctic seals proper, the ice-breeders, have not been subject to any real hunting. Since 1978 the ice-breeding seals have been subject to an international conservation convention which has set extremely low permissible catch limits in relation to the size of populations. There has been no commercial hunting of Antarctic seals since the 1978 convention.

Under the Convention for the Conservation of Antarctic Seals (CCAS), SCAR has been given a range of responsibilities, a move that is unique for a non-governmental body. Specifically, SCAR is invited to assess information; to encourage the exchange of scientific data and information; to recommend research programmes; to recommend data to be collected by sealing expeditions and to suggest suitable amendments to the Annex to the CCAS.

SCAR has also been invited to monitor the status of seal populations and to report when harvesting seals is having a significantly harmful effect on seals or the ecological system in any locality. The Group of Specialists on Seals, which developed from a subcommittee of the SCAR's Working Group on Biology, has this task. Through SCAR, the Group of Specialists would, in the unlikely event of commercial sealing starting in the Antarctic, provide advice to the participating governments of CCAS in order to govern the exploitation. Certainly this would last until the convention established its own scientific advisory committee.

WHALES

Whales as such do not form part of the main research purview of SCAR, but rather that of the International Whaling Commission. However, because many species feed on krill and thus play a part in the great Antarctic food web, whales have a place in this book.

Most species of great whales have counterparts in the Northern and Southern Hemispheres but their seasonal migrations are six months out of phase. The baleen, or filter-feeding, whales migrate from subtropical waters into the Antarctic where most of them spend about 120 days feeding, during which time they build up their blubber—in some species by 50 per cent. The blubber acts as an energy reserve for the rest of the year, because the whales take little food in subtropical waters. In Antarctic waters, they feed mainly on krill of varying size, but some species take significant amounts of small copepods (crustaceans resembling 'water-fleas').

The whales tend to follow the krill southwards as the Antarctic ice retreats in the summer. As the krill beneath the ice become exposed, the whales graze them and the northern boundary of the krill zone is thus gradually pushed southwards. The huge blue whales, which could be more than 30 m long, penetrate right into the pack ice. The fin whale (up to 26 m) is the next most southerly, and then species such as the humpback and the sei (both about 15 m long). The minke whale (up to 10 m long) is something of an anomaly because it is much smaller than the others, shows no big seasonal increase in its blubber layer and many

Minke whale

remain in the Antarctic over winter. The right whale (up to 18 m long) is a subantarctic species that feeds mainly on copepods at the Antarctic Convergence.

Of the toothed whales, the largest is the sperm whale (around 18 m) which feeds on squid. It develops a social structure similar to elephants— family groups consist of females and their calves, usually with a harem bull and some associated bachelors. With very few exceptions, only sexually mature bull sperm whales penetrate into Antarctic water. The sperm whales migrate seasonally and feed on the squid and fish around the Antarctic Convergence where they may be in competition with elephant seals which also feed on squid and fish.

There are about eight smaller toothed whales of which the killer whale is the most conspicuous. No accurate information is available on the population of dolphins but they are believed to feed on squid for the most part and not krill.

COMPETITION

So all the vertebrates of the Antarctic are linked within a great food web, directly or indirectly dependent on krill (see Fig. 12.1).

The baleen whales have been greatly reduced in number and are in competition for krill with the penguins and crabeater seals. The birds as a whole are now taking about 130 million tonnes of krill a year, while the seals are taking about 140 million tonnes.

Formerly the whales took about 190 million tonnes but that fell to an estimated 43 million tonnes. However, scientists think that the whales' share has probably increased as some of the species—notably the minke whale—no longer have to compete as previously for their food.

Thus from having been the major vertebrate consumer of krill, whales are now in third place. The consequence of this for the whales is that populations with the fastest turnover, such as minke, which has an annual breeding cycle and a 98 per cent pregnancy rate, as opposed to a two-year cycle and a 50 per cent pregnancy rate for the blue whale, have a competitive advantage. Of the seals, the crabeater has a 94 per cent pregnancy rate. It starts breeding at between two and three years. There is some evidence, although a little controversial, that the age of maturity of the crabeater seal and of some of the larger baleen whales has decreased in response to an increased growth rate. Current research is throwing some doubts on the exact meaning of this—it may be that the advancement of the age of maturity is not as great as the data appear to show. Of course, some penguins with their high breeding success (some have more than one fledged chick a year), a fairly long life-span and high consumption of food, also compete with the baleen whales. What Antarctic biologists now need to find out is how the different feeding ranges relate geographically because there is probably some ecological separation—even among the krill feeders—between those that are feeding on krill in the open ocean, north of the ice edge, and those that feed on krill at the ice edge, or in the pack ice zone. The major uncertainty is what happens in the pack ice zone in winter; ships seldom penetrate this zone then, and very little research has been done.

Information is not available to say which of the fluctuations in Antarctic living stocks are due to human activities and which to natural events. The krill population itself may fluctuate naturally responding to changed oceanographic conditions on a short or a long cycle. With the population of crabeater seals a clear four- to five-year cycle has been found and this is most unexpected in a large, long-lived mammal. It is difficult to interpret the cause of the cycle. It might be a response to fluctuations in krill abundance or availability, or a response to changes in the pack ice. But then again, it might be the result of a predator/prey relationship with the leopard seal: some Australian work on Macquarie Island has shown that leopard seals fluctuate with a periodicity almost identical to that of crabeater (4.7 years). The hint behind this is that the krill stock may have a similar short-term fluctuation, possibly imposed on a longer-term change over decades or even centuries.

LIFE ON THE SEA BOTTOM

The forms of life that live on the sea bottom (benthos) around Antarctica, especially that of the shallow shelf, have a biomass several orders of magnitude higher than that of the Arctic Ocean. In shallow waters there is an often dense multi-storied community of filter-feeding invertebrates. Some of the highest densities that have been recorded of invertebrates living in the bottom sediments have been found in shallow waters in the region of McMurdo Sound.

In contrast to earlier work, recent benthic studies have been quantitative. These studies have confirmed the widespread occurrence of characteristics such as high biomass, gigantism, a high degree of local peculiarities (endemism), an incomplete range of invertebrate groups, for example few barnacles or crabs, the relative absence of near-surface-living larval stages (normally considered to be agents of dispersal and colonization), high species diversity, high abundance, prolonged longevity, slow growth rates, and delayed maturation.

An exciting development concerns information on life beneath the ice shelves, provided by the Ross Ice Shelf Project (RISP). At a site more than 400 km from the open sea, the existence of life on the sea floor was confirmed. There were, however, low bacterial densities, sparse Crustacea at or near the bottom and no benthic forms in the sediments. The basal metabolism of the organisms was low.

PHYSIOLOGICAL STUDIES

A range of physiological studies has been undertaken on Antarctic invertebrates and fishes in particular. Cold-blooded animals in the Antarctic exhibit a number of features such as slow and seasonal growth, delayed maturation, longevity, large size, low fecundity, large egg size, non-pelagic larval development, seasonal reproduction and low metabolic rate, that appear to be an associated suite of mechanisms by which the organisms respond to cold, highly seasonal environments where primary production is confined to a brief period during the summer.

Recent work has challenged the original concept of cold-adapted metabolism (that is to say, elevated metabolic rates when compared with similar species from temperate environments) of Antarctic cold-blooded animals. Instead, there appears to be an overall reduction in their use of energy, which results in reduced basal metabolism, reduced growth and reduced reproductive effort in many species, while others such as krill live a very active life, at least in summer.

Considerable advances have also been made in our understanding of the ways in which fishes cope with temperatures approaching the freezing point of seawater, such as the evolution of unique antifreeze properties, and metabolic adaptation at the enzyme level. Studies of the physiology of white-bloodedness in the ice-fishes has determined how they extract sufficient oxygen from the water and transport it around their bodies.

TERRESTRIAL BIOLOGY

Beyond the coastal birds and seals, the native land animals of Antarctica are essentially limited to one wingless fly (*Belgica antarctica*)—a midge confined to the Antarctic Peninsula—various species of Collembola (springtails), and mites (some of which manage to tolerate $-60°C$ or colder). Two species of flowering plant are confined to the Antarctic Peninsula, the grass *Deschampsia antarctica* and the cushion plant *Colobanthus quitensis*. The non-flowering plants include lichens, mosses, algae, such as *Prasiola* and *Nostoc* (prominent in lakes), and microscopic soil fungi. A remarkable flora of blue-green algae and other micro-organisms inhabit the near-surface zone within some rocks.

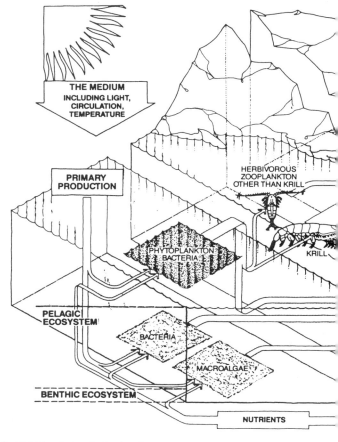

Fig 12.1 The Antarctic food web

Systematic botanical collections from continental Antarctica, throughout the maritime Antarctic and subantarctic islands (especially South Georgia, Macquarie and Marion Islands) date largely from 1960, and from these collections floras have been compiled and the associations between different species described. Such work is still being added to when the opportunity arises. The findings have encouraged subsequent ecological studies, notably a significant contribution to the International Biological Programme's Tundra Biome Studies and further research. The fauna has also been studied in increasing detail, in terms of densities, life cycle, behaviour, biochemistry and physiology, and energetics. Attention has been given to the processes which allow organisms to survive in the rigorous environment, for example, mechanisms of supercooling in the

Antarctic hair grass on Andrée Island, Antarctic Peninsula

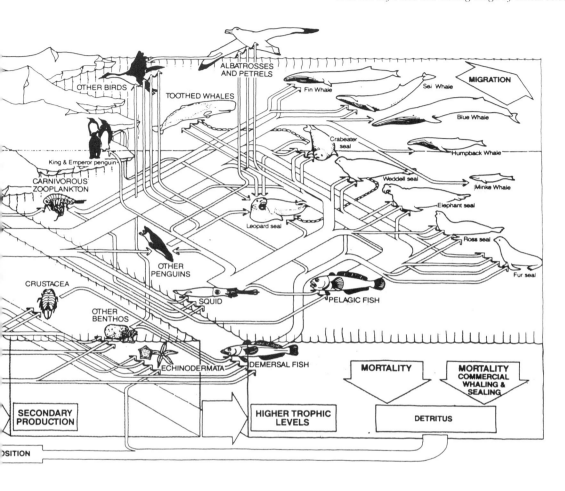

Antarctic pearlwort, with ripe seed capsules, on Andrée Island

insects, photosynthesis at low temperatures and humidities in the plants, and breeding strategies. Studies of how plants and animals survive sub-zero temperatures will continue, with special emphasis on the underlying pathways.

The relative simplicity of Antarctic terrestrial ecosystems, even compared with those of the Arctic, makes them ideal natural laboratories in which to test ecological and evolutionary hypotheses and progress towards the development of quantitative models of wider application. For example, the lack of indigenous large herbivores in the Antarctic and subantarctic, and the restricted distribution and diversity of introduced ones, provide unique opportunities for research aimed at understanding the roles of detritivores—the species of organisms, some microbial, some fungal, responsible for decomposing waste and

Pressing specimens of seaweed, Anvers Island, Antarctic Peninsula

dead matter in terrestrial ecosystems. There are also good opportunities for studying the ways in which new habitats become colonized, the sequence of events as various new colonies take over, characterizing successful colonizing attributes and relationships, and determining the significance of water for life at low temperatures. The emphasis will continue to be on studies of ecological processes where results should add not only to regional understanding but also to more general ecological theory of global significance.

Investigations of Antarctic soils are fundamental to an understanding of the genesis of temperate soils. Such studies include mechanical, chemical and biological weathering of rocks, the genesis of soil minerals at low temperature, geochemical cycling in systems mainly devoid of

humus, and activities linked to glaciation and its relationship to soil structure. These studies provide information on the rate at which soils form after glacial retreat.

Another significant area for further study is the interrelationship between the terrestrial and inland water ecosystems, which are poor in species and low in productivity, and the ecosystems of the surrounding oceans, which are relatively rich in species. The Antarctic seas are, as we have seen, the food source for huge populations of seabirds and seals which transport essential nutrients from sea to land. Knowledge of how much nutrient is transferred in this way is important, as also is that of how much material is flushed seaward from the terrestrial and inland waters.

ECOLOGY OF INLAND WATERS

The lakes and ponds of the Antarctic are among the most limited ecosystems found on the continent. The fastest growing areas of research have been in microbial ecology and in the chemistry and physics of natural waters. Emphasis is now on the interaction of organisms within the ecosystem, although the study of organisms in isolation is still important.

Conditions within the freshwater lakes are severe and they remain ice-covered for 8–12 months of the year, during which time they are virtually isolated from external influences and form ideal natural laboratories. Advantage is being taken of these special characteristics to study geochemical and biological processes of funda-

Algae and aquatic moss in a permanently ice-covered Antarctic lake

Freshwater lakes on Signy Island, South Orkney Islands

mental importance to modern limnological research and lake management schemes in other areas of the world, not just the Antarctic. Elsewhere such processes are difficult to measure because they occur at much higher rates and respond rapidly to more variables.

The Antarctic lakes are as yet unaffected by human beings, and they provide the opportunity for developing predictive models of unperturbed systems. The majority of the world's lake systems are virtually inaccessible. Antarctic lakes, on the other hand, occur around the fringe of the continent and there are many readily accessible examples for which data are now becoming available.

These data will also form a baseline for monitoring long-term global changes, for example acid rain, the accumulation of carbon dioxide, and local changes, for example, those due to mineral exploitation. Lakes are, in fact, natural 'sumps' of large catchment areas. Any modification of the catchment will be reflected in the lake system, which is more easily studied than the catchment itself.

13

Hostile environment for mankind

As this book has shown, Antarctica is the most isolated area on Earth and presents an extremely cold climate. The hostile environment is the great challenge to Man.

Mankind thrives in areas of the Earth that are best suited to support large numbers of other living organisms. Most favourable are the expansive plains and flood plains in middle latitudes and subtropics, and a few areas in the tropics. Through physiological mechanisms controlling sweating, blood pressure, ventilation, and skin permeability and colour, the human body has become adapted to a comparatively wide range of temperatures, atmospheric pressures, humidities, oxygen levels, and extremes of solar radiation. Despite all this, "clothing"—the adaptation that Man has taken on himself—provides him with the greatest protection against climatic extremes.

Although studies on human biology were undertaken on early Antarctic expeditions, they were *ad hoc* and had little continuity or co-ordination. The advent of SCAR and the co-ordination of international programmes by subsidiary groups, such as the Working Group on Human Biology and Medicine, have had a great impact on these studies. Human biological studies in Antarctica are planned so that scientists in all disciplines, can carry out their research more safely and efficiently in the hostile Antarctic environment.

Since the voyage of James Cook in 1775 there has been a long and interesting history of 'Antarctic' medical practice. However, the significance of human biological research was not realized until somewhat later than other scientific disciplines. As Antarctic environmental factors (isolation, cold and photoperiodicity) are not totally reproducible in the laboratory, an understanding of the effects of the Antarctic environment on Man and the hazards of Antarctic operations is central to a successful extended human involvement in the area.

Testing the ice on landing

ENERGY INPUTS

Normal energy needs in polar regions are the same as in temperate and tropical zones—between 12 500 and 15 000 kilojoules per day for average activity. However, the extra weight of heavy clothing and the muscular work of shivering increase caloric requirements by between 2 and 5 per cent. Heavy work may need as much as 25 000 kilojoules per day.

Certainly, adequate caloric intake is critical. Yet even when enough joules are provided, some people working in polar regions lose fat and increase muscle due to the conditioning effects of exercise. Some loss of weight, though, may be due to dehydration. With long-term stays in the Antarctic, blood pressure and the pulse rate may change. In parts of the body, the skin may thicken. Sometimes people experience a change in the quality of sleep, with a reduction of the rapid-eye-movement sleep associated with dreaming.

Since 1954, a substantial body of knowledge has been gained from human biological studies in Antarctica. These have been made at both permanent bases or stations and temporary field locations on the coast, and on the plateau at high altitude. While some of the research is directly applicable to an expedition or relevant to the health care of a single group, scientists have also undertaken unique basic studies in the Antarctic.

The range of topics is wide, including physical fitness, biochemistry and hormone studies, biorhythms, biometeorology, acclimatization to the cold, epidemiology, immunology, microbiology, virology, sleep, and the pathophysiology of cold injury. Studies on nutrition, clothing, buildings, and ergonomics have helped to produce better living and working conditions. Research on the psychology and behavioural adaptation of wintering groups has assisted in selection of personnel.

PROBLEMS OF THE COLD ENVIRONMENT

Acclimatization of Man to cold environment is a controversial issue. Anecdotal evidence suggests that it does occur to some extent. Hands probably become adapted to cold exposure, and this is manifest by less numbness and clumsiness, and by higher skin temperature.

Exposure to the cold may cause joint stiffness and the inability of muscles to contract normally. In frost-bite and other forms of freezing cold injury, damage to the local blood system may be severe: the outcome varies from complete healing to gangrene. Hypothermia—cold injury to the whole-body system—is caused by accidental environmental exposure to very low temperature.

Clothing can retard heat loss, but both moisture and exercise (because of the bellow-like action of clothes) decrease its insulative value. In cold

Entrance to a former Halley Bay Station, where the laboratory and living accommodation were 30 ft (over 9 m) below the surface

Amundsen-Scott South Pole Station. The station's geodesic dome houses three buildings for living quarters and work, 'skylab' and meteorological balloon-launching building

A communication tunnel at a research Station

water, exercise accelerates heat loss, while in cold air and warm water exercise may help to maintain body temperature.

Immediately following cold exposure, the pulse, the blood pressure, cardiac output, and stroke volume increase. Treatment of hypothermia is still controversial. Re-warming of a hypothermic victim depends on factors such as severity and duration and facilities at the treatment site. Careful maintenance and support of cardiovascular and pulmonary functions are desirable, but may not be possible in the Antarctic. Because certain disorders (primarily of the endocrine and neurological systems) can make people more susceptible to hypothermia, it is important that personnel for Antarctic work are screened for such conditions.

TRENDS IN ANTARCTIC HUMAN BIOLOGY

Human studies were part of the Norwegian–British–Swedish Antarctic Expedition of 1949–52. The foundations of polar physiology, however, were laid on the British North Greenland Expedition of 1952–54.

The setting up of SCAR in 1958 and the intensive planning for the International Biological Programme (IBP) in the 1960s brought together a group of interested medical workers as a subcommittee of the SCAR Working Group on Biology. They recommended a variety of standardized techniques and measurements to be made on Antarctic personnel and attempted to co-ordinate Antarctic studies of human biology.

In 1962, a meeting held in Paris, organized by the SCAR Working Group on Biology included papers dealing with human aspects. The first international gathering specifically on the special needs of humans in polar regions was also held in 1962. This was the Conference on Medicine and Public Health in the Arctic and Antarctic, organized by the World Health Organization in Geneva.

By the middle 1960s, most nations involved in the Antarctic had national programmes in human biology, but there was little collaborative work, although good lines of communication existed between the various members of the SCAR subcommittee. Following a SCAR/IUPS/IUBS symposium on 'Human Biology and Medicine in the Antarctic', held in Cambridge in 1972, the subcommittee was transformed into a full SCAR Working Group on Human Biology and Medicine.

THE FIRST BIOMEDICAL EXPEDITION

Since the inception of the working group, human biological research on an international co-operative level has improved. After four years of planning and organization, the International Biomedical Expedition to the Antarctic (IBEA) was successfully carried out in the austral summer 1980–81. Twelve scientists from five countries (Argentina, Australia, France, New Zealand and United Kingdom) participated in this the first Antarctic expedition solely for human biological studies. The aims of the expedition were to increase knowledge of Man's reaction to the Antarctic, and to compare the responses of artificially acclimatized and unacclimatized men working under field conditions in Antarctica.

The IBEA research programme was multidisciplinary with projects in physiology, biochemistry, microbiology, immunology, psychology and behavioural adaptation, sleep, and epidemiology. After initial experiments in Australia, the group spent 10 weeks travelling by motorized toboggans and living in tents on the Antarctic plateau. At a workshop in Paris in January 1983 all participants were brought together for preliminary discussion of their work, and the results were presented at the Sixth International Symposium on Circumpolar Health, held in Alaska, in 1984.

Frequent exchanges of data and information, by correspondence and meetings, have led to improved health care services and also a greater activity in, and level of, research. Interest has spread beyond the SCAR Working Group itself to other agencies including the WHO, the Nordic Council for Arctic Medical Research, and the newly established International Society for Circumpolar Health.

CURRENT AND FUTURE RESEARCH AIMS

Current research continues a multidisciplinary approach to understanding the interaction of Man with the Antarctic environment. Recent research places particular emphasis on studies that facilitate living in Antarctica. Increases in the number of visitors, and of private and national expeditions, mean increasing populations. The advent of families and increasing numbers of women have changed the traditional all-male groups. These changes will require continuing studies, many of which will be applied.

So far as medical research is concerned, there are no known polar-peculiar illnesses. However, more research is needed on the immune system, circadian rhythm, wound healing, cold injury, and the effects of changing daylength and electromagnetic fields. Some SCAR nations want to see more research into the possible impact of varying lengths of residence in Antarctica. At present, little is known of the health risks, if any, for an individual who has spent much of his or

The long 'days' of winter darkness begin to draw in

Diving through winter ice near Syowa Station— physiologists would like to know more about cold stress incurred by working in Antarctic waters

her life in the Antarctic. Other research topics include the effect of dehydration on both physical and mental performances during polar travel, and the cold stress of diving in Antarctic waters.

Through international scientific collaboration, it should be possible to maintain research at a level of high quality, while also giving opportunity for cross-cultural studies. Work on IBEA has shown that physiological, microbiological and immunological research can be greatly helped by parallel psychological studies. Similarly, psychological and social data from small, isolated Antarctic groups indicate a need for more sophisticated research in this area to include biological factors. In some quarters, it has been suggested that Antarctic medical studies would be

Preparing to dive through summer ice

Biological research under the ice near Syowa

more appropriately done in the Arctic, where indigenous Inuit people have lived in family communities for millennia. At the present time, groups from temperate zones live in military, and oil and mineral exploration establishments in the Arctic. However, these groups are not totally isolated, as staff have frequent opportunities to travel to recreational centres. As well as there being chronic and endemic disease in the Arctic, the existence of indigenous and non-native groups creates problems between the two cultures. For these reasons, much of the research done in Antarctica cannot be done in the Arctic.

The isolation of Antarctica is an ideal situation for the study of viral, immunological, bacteriological and psychological research. In the age of space exploration and the current evaluation of Man's impact on the world and of the world on himself, human biological studies in the Antarctic are highly relevant.

14

Environmental impact of Man

The exploitation of the Antarctic and its resources is possible in the future. Partly with this in mind, SCAR biologists and conservationists are now planning research programmes to identify vulnerable ecosystems or habitats, to describe their dynamics and to determine background levels of contaminants. The marine ecosystem has already been grossly disturbed by Man's exploitation of whales and seals. However, Man's impact on the terrestrial and inland water ecosystems is slight, as we have seen. Study of the presently undisturbed situation and the inherent variability is important to provide an initial quantitative description against which the consequences of possible future increased human activity can be assessed. Much of this research is of a long-term nature.

Scientists engaged on Antarctic research and their supporting organizations are aware of the need to take special care that their research and logistic activities in no way needlessly interfere with Antarctic ecosystems. In Canberra, in 1983, the Twelfth Antarctic Treaty Consultative Meeting (ATCM) called for an assessment of the impact of mankind on the Antarctic environment. As a first step, the meeting asked SCAR to advise on the types of research and systems of supply that might affect the Antarctic environment. Furthermore, the Canberra meeting asked SCAR to suggest ways of assessing any environmental changes due to human activity on the continent.

SCAR responded to these requests by producing a report* that presents its philosophy and criteria for assessing the status of the unique Antarctic environment in which mankind *directly* plays only a very minor role. Assessments and monitoring of human activities in the Antarctic, recommends the SCAR report, should go beyond their biological effects and onto the changes that they cause in the physical and chemical environment of the continent. To be meaningful, such assessments must consider not only how fragile food chains are affected by Man's activities, but also any untoward changes that such activities

Man's Impact on the Antarctic Environment, by W. S. Benninghoff of the University of Michigan and W. N. Bonner of the British Antarctic Survey (SCAR 1985).

Elephant seals rest besides whale bones, wastes of former exploitation

cause in the vast ice sheet and the atmosphere, particularly the stratosphere.

'Antarctic operators should prepare environmental assessments for their research stations and the major recurrent operations connected with the stations.' Moreover, such assessments 'should cover both the initial acute impacts and the subsequent operational impacts. If major operational impacts are identified, the preparation of a full environmental impact statement should be considered. In either case, a systematic review of the impacts that have occurred or are still taking place might suggest ways of reducing or eliminating them.'

The report emphasizes that, because human activity in the Antarctic is growing, assessment of the Antarctic environment should be a continuing process. Environmental impact assessments prepared for research stations and recurrent operations should be reviewed at regular intervals. In order to detect unforeseen operational impacts, a monitoring system is required. The choice and siting of monitoring stations and the frequency of measurements will depend on what is being monitored and where.

JUST A SCRATCH ON THE SURFACE?

The greatest value of the Antarctic continent to mankind resides in the wealth of information it contains and yields about planet Earth: global weather systems both past and present, the geological history of the Southern Hemisphere continents, the structure and dynamics of the magnetic envelope around the Earth, the influence of solar radiation on the atmosphere, and the remarkable adaptations of organisms to the extremes of cold and isolation.

Environmental protection in the Antarctic has two components. The first is concerned with the maintenance of the high productivity and the ecological relationships in the Southern Ocean. The second involves maintaining the unspoiled environment and the fragile ecosystems of the land, inland waters, and ice-covered areas of the continent.

IMPACTS ON MARINE AND TERRESTRIAL ECOSYSTEMS

The Southern Ocean is an immense dynamic system with relatively constant currents, with a regular seasonal advance and retreat of sea ice over an area greater than the continent itself, and also the regular input of icebergs from ice shelves. The open ocean provides mixing, diffusion and buffering of added substances that are incorporated relatively rapidly into the food chains and into the sea-bed sediments. Inshore areas, especially sheltered bays, do not have such strong aids to dispersion or such a large biomass to assimilate additives. As a consequence, these areas are more liable to be changed by the intrusion of pollutants. Recovery of the marine ecosystem from an oil spill, for example, could take more than a decade in an enclosed bay, whereas in the open sea the spill would probably not be detectable after a year.

On the ice-free surfaces of the continent, there is generally a thin scattering of minute plants and animals—lichens, algae, fungi and invertebrates including nematode worms, mites, and springtails. In this polar desert, the few species of plants and animals can only function when the snow cover melts, or the soil warms enough to yield films of liquid water. The terrestrial biological processes operate only slowly and intermittently, and on very small scales. Consequently, the ecosystems are easily disrupted or perturbed, and recover only slowly. Lichens, if disrupted, could take several centuries to recover.

In general, the terrestrial ecosystem and, in particular, the freshwater bodies within it, are fragile, having little capacity to withstand chemical, physical, or biological changes without themselves being changed.

EXPEDITION BASES AND RESEARCH STATIONS

Expedition bases and stations are often sited adjacent to the coast on areas of level rock, where they are free from permanent snow and ice, and where the pack ice retreats in early summer. The same features that are attractive to human beings also favour the establishment of biological communities, for example, penguin rookeries. For this reason, many research stations have a high probability of causing impacts. Certainly, all research stations in Antarctica need electrical power and heat, whose generation leads to the emission of gases, waste heat, dust, and noise. In the past, nuclear-powered generators were sometimes advocated because it was generally believed that they would, in principle, cause less environmental impact than would conventional oil-powered

Abandoned waste from stations, such as this, are a thing of the past. Most of these unsightly dumps have now been cleared up and removed from the continent

Rubbish, mostly from fishing boats, continues to arrive on Antarctic beaches

systems. Only *one* nuclear power station has ever been used in Antarctica, but that has now been removed.

So far as effluent and waste treatment is concerned, care must be taken that the technology used does not create a greater impact than the wastes themselves. Ships can and should be used to transport waste from the continent: the expense incurred can be justified on the grounds that the Antarctic's near-pristine condition demands special treatment: waste recycling must be done elsewhere.

Antarctic research stations can offer a route and a reservoir for organisms alien to the continent. The introduction of micro-organisms with the potential to influence Antarctic living systems is a threat most likely to originate at research stations where infected domestic animals (for example, caged birds) or exotic soils may arrive on ships or vehicles.

A RANGE OF OTHER INFLUENCES

Among the other categories of activity that might cause a 'significant impact' on the Antarctic environment are: airstrips, increases in research personnel and tourism, increases in fuel consumption, operations that might affect specially protected areas or sites of special scientific interest, perturbation experiments extending over more than 100 m^2, drilling operations, marine seismic surveys involving the use of explosive charges, and the introduction of radionuclides into the environment where their recovery and removal cannot be assured.

Since the advent of nuclear explosions and nuclear power plants elsewhere in the world, levels of krypton-85 in the atmosphere have increased. The presence of this isotope in the atmosphere has the effect of increasing the electrical conductivity of the air. The phenomenon needs further investigation and monitoring to determine the potential effects that this isotope is having on radio transmission and on electrostatic forces active in blowing snow and dust.

Scientists have found the insecticide DDT in penguin fat and eggs, thus indicating that this chemical can become transported over very long distances through marine food chains.

The SCAR report comments that given 'the diversity of physical, chemical, and biological elements operating in the environment, there must be continuing search for the "pressure points" at which it is possible to detect changes of a kind or class that require consideration in an environmental impact assessment'.

SPECIAL CONSERVATION AREAS

SCAR published in mid-1985 an atlas of *Conservation Areas in the Antarctic*. The Antarctic Treaty recognized the importance of conservation as long ago as 1964. The 'Agreed Measures' designed by the Antarctic Treaty Consultative Meeting (ATCM) cover plants, birds, and sea mammals (lower forms of life are not yet pro-

Tourists photographing a Weddell seal: non-scientific visits are on the increase

Antarctic scientists take care to cause minimal disturbance to the environment

tected) throughout the whole of the land and ice shelves south of latitude 60° South. It is impracticable to designate parts of the open sea as conservation sites. To achieve that, other more general agreements are required.

In 1982, the Subcommittee on Conservation invited the national committees of the various member nations of SCAR to propose extensions to boundaries of existing conservation sites, and to identify other areas for special attention. SCAR's annotated atlas sets out existing and proposed Specially Protected Areas (SPAs) and Sites of Special Scientific Interest (SSSIs) and will be updated periodically as new SPAs and SSSIs are nominated. The ultimate aim of the SPA and SSSI network is to ensure that sufficient representational and different geographical areas of the Antarctic and the subantarctic regions are included. The subcommittee devised a system of classification of marine, terrestrial and inland ecosystems to ensure that all types of Antarctic habitat are adequately represented and preserved.

The designation of an SPA virtually isolates it from active scientific research programmes: 20 SPAs exist under the Agreed Measures of the Antarctic Treaty. However, there is a good case for giving sites a measure of protection while research is in progress. With this in mind, 21 SSSIs have already been established.

Outside the Antarctic Treaty area, north of 60° South, there are nine islands considered to merit special protection because of their relevance to the Antarctic.

Although the Antarctic Treaty has no control over the subantarctic islands north of latitude 60° South, they come within the purview of SCAR.

Even with few visitors and scientists, in time tracks appear, as here on a hillside in Moraine Valley, Signy Island

Scientific site at Moraine Valley, Signy Island, South Orkney Islands

15

Freedom of exchange of Antarctic scientific information

The history of the free exchange of scientific information is a long one that predates the first International Polar Year (IPY) (1882–83). Standard formats for recording observations began to be introduced in the early 18th century in astronomy and meteorology. In the 19th century, Carl Gauss and Wilhelm Weber, working in the framework of the Magnetic Union, introduced not only standard formats but also standard times for observations of magnetic phenomena. Gauss also attempted to introduce standard instruments but with less success.

Karl Weyprecht's first proposal for a ring of circumpolar stations—which became the first IPY—included a plan for the collection and publication of geophysical observations made at standard times. He also suggested the use of standard instruments. However, as this proved impossible, many of the instruments taken with the expeditions were compared and calibrated with standard observatory instruments, both before departure and after return, in order to ensure maximum compatibility and comparability of the observations.

The tradition of free exchange of information and data continued when the first observations were made on the Antarctic continent. To the Belgian Expedition on the *Belgica*, led by Adrien de Gerlache in 1897, goes credit for the first long-term, continuous series of observations in the Southern Ocean and on the Antarctic ice. These were published in the years immediately after the *Belgica*'s return.

During the second IPY (1932–33) a standard data centre was established in Copenhagen and some observations* were published centrally. The experience of some of the participants of the second IPY, who were involved in the planning of the IGY, was certainly responsible for the empha-

sis laid on the collection of standard observations in more than one World Data Centre. The amount of information available from Antarctica and the Southern Ocean changed dramatically with the IGY when 12 countries created 44 new research bases in addition to the 11 that already existed. Not only did the amount of information increase but so did the rapidity with which some of it was made available. Whereas for the earlier expeditions the observations were published after the return of the expeditions, in the IGY arrangements were made to transmit in real time some meteorological observations. Such data were of importance in improving weather forecasts for the Southern Hemisphere.

Other observations were transmitted, according to procedures agreed by the nations participating in the IGY, to the IGY World Data Centres—of which there were two principal ones (WDC A in the United States and WDC B in the Soviet Union) for each of the major IGY disciplines, with smaller WDC's C elsewhere which specialized in one aspect of data. The WDCs have been continued under the auspices of ICSU and now form an important and continuing element of international collaboration in science. The centres agreed to ensure that the scientific information they contained would be readily available on request at the cost of reproduction to scientists throughout the world. Special regulations ensure that due acknowledgement is made to the use of the data and that priority is given to the scientists who made the observations.

It was natural, therefore, that when SCAR was created the constitution included:

'SCAR will encourage and assist in the dissemination of scientific knowledge derived from research carried out in the Antarctic.'

SCAR also introduced the principle of National Reports so as to ensure regular exchange of infor-

---

* Some were lost during the 1939–45 war.

mation on scientific programmes and plans between SCAR nations, and to ensure that specialists could be aware of all investigations made in the Antarctic in their particular disciplines. In addition, SCAR's working groups and groups of specialists are required to establish and to maintain suitable links between each other and with relevant associations or committees of ICSU.

At the first Antarctic Treaty Consultative Meeting in July 1961, the tradition of free exchange of information was included in recommendation I-IV, which states: 'that the free exchange of information and views among scientists participating in SCAR, and the recommendations concerning scientific programmes and co-operation formulated by this body, constitute a most valuable contribution to international scientific co-operation in Antarctica.'

In recent years, there has been a marked increase in output from Antarctic science. This is partly due to the increase in the number of Antarctic scientists since the IGY but there is also the increasing influence of technological improvements. Observational sciences, whether biological, geological, or physical, can produce very large sets of raw measurements. The volume will grow with the increasing use of electronic automatic methods of gathering data. Satellite-sensing techniques provide voluminous quantities of automatic data. Likewise, the new advanced ionospheric sounders produce 3500 megabytes of data per year (that is equivalent to a 700-million-word book each year). Automatic equipment can also extend surface-observing networks by operating in all weather and on terrain where human observers would find it difficult to work.

Research material brought back from the Antarctic can be stored in many forms. Sample collections form the basis of identification work particularly in biology, geology, and some glaciology. These collections are usually held by the country that collected them, and they are often maintained at a single site in that country.

Sometimes collections are very extensive: the plankton specimens from the voyages of the *Discovery II*, during the 1930s, are still being analysed. The *Discovery Reports* themselves now run to 37 volumes.

The availability of such collections to scientists of other countries depends on the attitude of the individual centre but generally, scientists working in the same area of research are welcome to examine such collections, especially when they include type specimens of a species.

Most results are made available in the normal way by publication of scientific papers but certain types of data are made available in compiled or analysed form, such as maps and charts. As an example of international exchange, the USSR collected and compiled gravity data from all Antarctic stations. Maps are usually produced to show geological strata or hydrographic soundings: however, since the IGY, enough data have been available to create contour maps that connect areas with a similar value (isolines) for some environmental variables, such as magnetic flux. In 1978, Japan produced a series of magnetic maps based on data from many SCAR countries. SCAR countries also contributed data for the preparation of the Antarctic sheet of the *General Bathymetric Chart of the Oceans*, Fifth Edition, sheet 18.

Environmental data from satellites are also often displayed using maps. LANDSAT images of Antarctica have provided the basis for planimetric mapping. The meteorological and oceanic satellites of the US National Oceanic and Atmospheric Administration (NOAA) have produced regular images from infra-red and radio waves emitted from natural surfaces. These have been used mainly for meteorological information, or for showing the extent of snow or ice cover (see for example NASA's *Antarctic Sea Ice 1973–1976: Satellite Passive-Microwave Observations*). Satellite images can also be used to infer observations. The colour scanners on the NIMBUS series can be used to assess abundance of phytoplankton and this technique has been used for the Southern Ocean.

Biological data are not normally included in the World Data Centres although the WDCs for Oceanography do contain some biological oceanographic information. One of the reasons for this is the great variety of biological data and the difficulties that have been encountered in providing standard formats for recording information in numerical, computerized, form. However, a number of special arrangements have been made for the international handling of such data from the Antarctic seas. The International Whaling Commission (IWC) has a computerized data bank at its headquarters near Cambridge, England, with information on whale catches, sightings, and some other biological data concerning whales going back to the 1940s.

As part of the effort to provide a centralized store of marine biological data, a new BIOMASS Data Centre was created in Cambridge, England, using information collected during the two BIOMASS experiments (see Chapter 12). Using experience gained from earlier biological analyses, the data collected by 11 countries have been entered onto a large data base system, under the auspices of the BIOMASS Executive. Although the data collection phases of the BIOMASS experiment have now finished, much of the data on krill and phytoplankton ecology have still to be analysed. Once validated, the eventual aim is to maintain an interactive database as a long-term resource for marine research, and to contribute to wider aspects of Antarctic data exchange. This is the first time that such extensive marine biological research data have been handled in this way. Information on Antarctic marine living resources is also being collected within the framework of the Convention for the Conservation of Antarctic Marine Living Resources (see Chapter 12).

Some information about fishery resources in the Southern Ocean has been compiled by the Food and Agriculture Organization (FAO) in its statistics on world fisheries. Also, FAO, in conjunction with the Intergovernmental Oceanographic Commission (IOC) and the UN Ocean Economics and Technology Branch (OETB) operates an Aquatic Sciences and Fisheries Information System (ASFIS), which is preparing a specialized referral service called the Marine Environmental Data and Information Referral System (MEDI).

Geological data also are not normally included in the world data centres but with the development of the International Geological Correlation Programme—organized jointly by the International Union of Geological Sciences and UNESCO—some of the problems of comparability and compatability of the information are being overcome. Information about the geology of the Antarctic, the subantarctic islands, and the floor of the Southern Ocean are published in the scientific literature and reported on at international scientific meetings.

SCAR keeps under regular review the level of exchange of data. The subsidiary groups of SCAR play a prominent part in ensuring the continued exchange of data. In 1983 a climate data workshop was held in which were highlighted some of the areas where the transfer of data to the WDC's was not complete, for example in glaciology and meteorology, and proposals were made to improve the situation.

The specialized symposia of SCAR, such as those on Antarctic Glaciology, Antarctic Geology, Antarctic Biology and the Polar Oceans, (co-sponsored by SCOR), provide excellent occasions for the exchange of information and an overall view of the state of science in a particular discipline. Similarly, the meetings of the SCAR working groups and of SCAR groups of specialists provide further opportunities for such exchanges and reviews. The reports of the meetings are published by SCAR and are freely available.

The meetings of many other bodies in the ICSU family facilitate the exchange of information and data about the Antarctic, and enable scientists interested in Antarctica to meet. Two recent meetings are given as examples: COSPAR/SCAR Workshop on Satellite Observations of the Antarctic: Past, Present and Future; ICPM Symposia on Polar Meteorology.

Finally, information about recent polar literature is published in a variety of sources, perhaps the best known being the US Cold Regions Bibliography Project, which publishes *Current Antarctic Literature* and the *Antarctic Bibliography*, (the latter contains 32 740 entries spanning 1951 to the end of 1985), and the Scott Polar Research Institute, which publishes *Recent Polar and Glaciological Literature* (it presently lists about 5000 entries a year). Most Antarctic institutes produce their own bulletins or reports, although these are often only easily available at, or from, the institute in question.

16

A continent for collaboration

The foregoing chapters should have made it clear that, particularly in the past three decades, a great amount of scientific research has been carried out in the Antarctic and that much remains to be done. It is of importance that mankind increases his understanding of the processes of nature in all parts of the planet on which he lives and his understanding of the influence exerted by himself and by each part on the other parts. Research has shown that significant interactions occur between polar and temperate regions, particularly manifest in studies of climatology and oceanography but also in other fields such as the biological and Earth sciences.

Results of scientific research in Antarctica and the surrounding ocean are freely available within the international scientific world—not only between scientists of countries themselves active in Antarctic research but to scientists of all other countries as well. Thousands of scientific papers resulting from research undertaken in Antarctica and the surrounding ocean have been published in the open scientific literature and some programmes contributed data directly to global programmes of research. The world's scientific community has long recognized the need for scientific research in Antarctica and the surrounding ocean. Few nations, however, have possessed the will as well as the sophisticated technologies and scientific resources necessary to undertake major programmes in the harsh and difficult regions of the Antarctic. Governments with adequate resources have made them available for research, with little material reward for themselves, as a contribution to Man's understanding of his global environment. The results benefit all peoples of the world, and the openly-available scientific knowledge gained has provided, and still does provide, the only factual basis on which governments can make rational judgements on emerging practical issues, such as those relating to the resources of the region.

The number of countries contributing to Antarctic research has grown in recent years and it is expected, and hoped, that this trend will continue. This is especially so now that, in most scientific disciplines, it can be said that the exploratory or reconnaissance phase is past and that major research problems are now emerging. Almost 50 scientific stations are now operating all the year round, but because of the vast size of the continent the network remains sparse. From these stations also operate field parties, supported by over-snow vehicles or aircraft. The unique nature of the Antarctic Treaty actively promotes the concept of freedom of scientific investigations and freedom for scientists to work anywhere they wish. This provides a sound governmental basis for scientists of different nations to work together. The scientists themselves are anxious to do this because of the magnitude of many of the problems needing to be studied, some of which are beyond the capability of any one nation to tackle alone. Also, in many instances, the pooling of resources by two or more nations for a particular study enables more effective uses to be made of these resources. This applies not only to work on the continent of Antarctica itself but also to research at sea. Great benefit is derived from multi-national teams of scientists on a research cruise, or when research ships of more than one country engage in a collaborative project.

Collaboration between scientists with different areas of expertise of different countries is, and always has been, the guiding spirit of SCAR with its multinational working groups, groups of specialists and programmes. The spirit of active collaboration between peoples of different nations in addressing issues of common interest is refreshing in the troubled world of today. It is important to all mankind to ensure that the existing goodwill and international collaboration among Antarctic scientists continue to be encouraged so as to permit the scientific research to advance in the future as it has since the conception of the International Geophysical Year.

Appendix 1: SCAR Publications

BIOLOGICAL

Biologie Antarctique/Antarctic Biology, edited by R. Carrick, M. Holdgate and J. Prevost: pp 651, 1964. Paris, Hermann [First SCAR/IUBS Symposium on Antarctic Biology, Paris 2–8 September 1962].

Antarctic Ecology, edited by M. W. Holdgate: pp 998, 2 vols, 1970, London [Proceedings of Second SCAR/IUBS/SCIBP Symposium on Antarctic Biology, Cambridge 29 July–3 August 1968].

Adaptations within Antarctic Ecosystems, edited by G. A. Llano: pp 1252, 1977. Washington, DC, Smithsonian Institution [Proceedings of Third SCAR/IUBS Symposium on Antarctic Biology, Washington 26–30 August 1974].

Antarctic Nutrient Cycles and Food Webs, edited by W. R. Siegfried, P. R. Condy and R. M. Laws: pp 700, 1985. Berlin Heidelberg New York Tokyo, Springer Verlag [Proceedings of the 4th SCAR, SCOR, IABO, SASCAR Symposium on Antarctic Biology, Wilderness, South Africa, 12–16 September 1983].

Conservation Areas in the Antarctic, edited by W. N. Bonner and R. I. Lewis Smith: pp 299, 1985. Cambridge, SCAR ISBN 0 948277 01 7.

Biological Investigations of Marine Antarctic Systems and Stocks (BIOMASS): Research Proposals (BIOMASS Scientific Series 1): pp 79, 1977. Cambridge, SCAR and SCOR.

Biological Investigations of Marine Antarctic Systems and Stocks (BIOMASS): Selected Contributions of the Woods Hole Conference on Living Resources of the Southern Ocean 1976 (BIOMASS Scientific Series 2), edited by S. Z. El-Sayed: pp 155, 1981. Cambridge, SCAR and SCOR [Selected contributions of the Conference on Living Resources of the Southern Ocean, Woods Hole 17–21 August 1976].

Swimming Behaviour, Swimming Performance and Energy Balance of Antarctic Krill, Euphausia superba (BIOMASS Scientific Series 3), by Uwe Kils: pp 122, 1981. College Station, Texas, SCAR and SCOR.

Distribution and Abundance of Antarctic and Sub-Antarctic Penguins: A Synthesis of Current Knowledge (BIOMASS Scientific Series 4), compiled by G. J. Wilson: pp 46, 1983. Cambridge, SCAR and SCOR.

Illustrated Guide to Fish Larvae of the Southern Ocean (BIOMASS Scientific Series 5), by F. N. Efremenko: pp 74, 1985. Cambridge.

Biology and Status of Exploited Antarctic Fish Stocks (BIOMASS Scientific Series 6), by K.-H. Kock, G. Duhamel and J.-C. Hureau: pp 143, 1985. Cambridge, SCAR and SCOR. ISBN 0 948277 03 0.

Polar Human Biology, edited by D. G. Edholm and E. K. E. Gunderson: pp 443, 1973. London, William Heinemann Medical Books [Proceedings of SCAR/IUPS/IUBS Symposium on Human Biology and Medicine in the Antarctic, Cambridge 19–21 September 1972].

Man in the Antarctic, edited by J. Rivolier, R. Goldsmith. D. J. Lugg, and A. J. W. Taylor: pp 250. London and Philadelphia, Taylor and Francis; in the press. [The scientific work of the International Biomedical Expedition to the Antarctic, IBEA].

EARTH SCIENCES

Antarctic Geology, by R. J. Adie: pp 758, 1964. Amsterdam, North-Holland Publishing Co [Proceedings of First SCAR/IUGS International Symposium, Cape Town 16–21 September 1963].

Antarctic Geology and Geophysics (IUGS Series B— No. 1), edited by R. J. Adie: pp 876, 1972. Oslo, Universitetsforlaget [Proceedings of SCAR/IUGS Symposium on Antarctic Geology and Solid Earth Geophysics, Oslo 6–15 August 1970].

Antarctic Geoscience (IUGS Series B—No. 4), edited by C. Craddock: pp 1172, 1982. Madison, The University of Wisconsin Press [Proceedings of SCAR/IUGS/ICG Symposium on Antarctic Geology and Geophysics, Madison 22–27 August 1977].

Antarctic Earth Science, edited by R. L. Oliver, P. R. James and J. B. Jago: pp 697, 1983. Canberra, Australian Academy of Science [Proceedings on Fourth SCAR/IUGS International Symposium on Antarctic Earth Sciences, Adelaide 16–20 August 1982].

Circum-Antarctic Marine Geology. In: *Marine Geology*, 25, No. 1/3 (Scientific Report No. 35), edited by D. E. Hayes: pp 277, 1977. Amsterdam, Elsevier Scientific Publishing Co [Proceedings of CMG/SCAR Symposium, Sydney 18–19 August 1976].

GLACIOLOGICAL

Colloque sur la Glaciologie Antarctique/Symposium on Antarctic Glaciology (IAHS Publication No. 55): pp 162, 1961. Gentbrugge, De l'Association Internationale d'Hydrologie Scientifique [Proceedings of the IAHS/SCAR Symposium, Helsinki, 25 July–6 August 1960].

International Symposium on Antarctic Glaciological Exploration (ISAGE) (IAHS Publication No. 86), edited by A. J. Gow, C. Keeler, C. C. Langway and W. F. Weeks: pp 543, 1970. Gentbrugge and Cambridge, International Association of Scientific Hydrology and SCAR [Proceedings of SCAR/IAHS Symposium, Hanover 3–7 September 1968].

Annals of Glaciology (vol. 3): pp 362, 1982. Cambridge, International Glaciological Society [Proceedings of Third International Symposium on Antarctic Glaciology, Columbus 7–12 September 1981].

OCEANOGRAPHY

Symposium on Antarctic Oceanography, edited by R. I. Currie: pp 268, 1966 [Proceedings of SCAR/SCOR/IAPO/IUBS Symposium, Santiago 13–16 September 1966].

Symposium on Antarctic Ice and Water Masses, edited by Sir George Deacon: pp 113, 1971. Cambridge, SCAR [Proceedings of SCAR Symposium, Tokyo 19 September 1970].

Polar Oceans, edited by M. J. Dunbar: pp 682, 1977. Calgary, Arctic Institute of North America [Proceedings of SCOR/SCAR Conference, Montreal May 1974].

Atlas of Polish Oceanographic Observations in Antarctic Waters 1981: pp 83, 1985. Cambridge, SCAR and SCOR (A BIOMASS incidental publication). ISBN 0 948277 04 1.

METEOROLOGY AND CLIMATOLOGY

Antarctic Meteorology, pp 483, 1967. Melbourne, Pergamon Press [Proceedings of Symposium, Melbourne 18–25 February 1959].

Polar Meteorology: WMO Technical Note No. 87 (WMO-No. 211 TP. 111): pp 540, 1967. Geneva, World Meteorological Organization [Proceedings of WMO/SCAR/ICPM Symposium, Geneva 5–9 September 1966].

Energy Fluxes over Polar Surfaces: WMO Technical Note No. 129 (WMO No. 361), edited by Svenn Orvig: pp 300, 1973. Geneva, World Meteorological Organisation [Proceedings of IAMAP/IAPSO/SCAR/WMO Symposium, Moscow 3–5 August 1971].

Antarctic Climate Research: Proposals for the Implementation of a Programme of Antarctic Research Contributing to the World Climate Research Programme, edited by I. Allison, pp 65, 1983. Cambridge, SCAR.

PALAEOECOLOGY

Palaeoecology of Africa and of the Surrounding Islands

and Antarctica (vol 5), edited by E. M. van Zinderen Bakker: pp 240, 1969. Cape Town, A. A. Balkema. [Proceedings of SCAR Conference on Antarctic Quaternary Studies, Cambridge 24–27 July 1968.]

Palaeoecology of Africa and the Surrounding Islands and Antarctica (vol 8), edited by E. M. van Zinderen Bakker: pp 198, 1973. Cape Town, A. A. Balkema [Proceedings of SCAR Conference on Antarctic Quaternary Studies, Canberra 9–12 August 1972].

Antarctic Glacial History and World Palaeoenvironments, edited by E. M. van Zinderen Bakker: pp 172, 1978. Rotterdam, A. A. Balkema [Proceedings of SCAR Symposium, Birmingham 17 August 1977].

MISCELLANEOUS

Man's Impact on the Antarctic Environment, by W. S. Benninghoff and W. N. Bonner: pp 56, 1985. Cambridge, SCAR. ISBN 0 948277 00 9.

Possible Environmental Effects of Mineral Exploration and Exploitation in Antarctica, edited by J. H. Zumberge: pp 59, 1979. Cambridge, SCAR [Adaptation of a report by the SCAR Group of Specialists on the Environmental Impact Assessment of Mineral Resource Exploration and Exploitation in Antarctica].

Antarctic Environmental Implications of Possible Mineral Exploration and Exploitation (AEIMEE), editor R. H. Rutford: pp 95, 1986. Cambridge SCAR ISBN 0948277–05. US $10 or £6.
SCAR has now published revised and slightly updated versions of the 1981 and 1983 reports of the SCAR Group of Specialists on Antarctic Environmental Implications of Possible Mineral Exploration and Exploitation. Also included in the volume are the various relevant Antarctic Treaty Consultative Meetings recommendations and reports: Order with remittances, to: SCAR, The Distribution Centre, Blackhorse Road, Letchworth, Herts SG6 1HN, England. Price includes unsealed airmail postage.

Journal of the Royal Society of New Zealand, vol 11, No. 4, pp 205, 1981. Wellington, Royal Society of New Zealand [Proceedings of the SCAR Symposium on the Ross Sea, 17–18 October 1980, Queenstown, New Zealand].

Antarctic Telecommunications, edited by A. H. Sheffield: pp 402, 1972. Cambridge, SCAR [Proceedings of SCAR Symposium on Technical and Scientific Problems affecting Antarctic Telecommunications, Sandefjord 10–16 May 1972].

Symposium on Antarctic Logistics, pp 778, 1963. Washington DC, National Academy of Sciences/National Research Council [Proceedings of SCAR Symposium, Boulder 13–17 August 1962].

Appendix 2: Antarctic Treaty

Made 1 December 1959; came into force 23 June 1961.

The Treaty has no limit on its duration; it may be reviewed, at the request of a Consultative Party, 30 years after coming into force [that is in 1991].

Signatory states; in chronological order.

§ *United Kingdom*	*31 May 1960*	1
§ *South Africa*	*21 June 1960*	2
§ *Belgium*	*26 July 1960*	3
§ *Japan*	*4 August 1960*	4
§ *United States of America*	*18 August 1960*	5
§ *Norway*	*24 August 1960*	6
§ *France*	*16 September 1960*	7
§ *New Zealand*	*1 November 1960*	8
§ *Soviet Union*	*2 November 1960*	9
§ Poland*	8 June 1961	10
§ *Argentina*	*23 June 1961*	11
§ *Australia*	*23 June 1961*	12
§ *Chile*	*23 June 1961*	13
Czechoslovakia	14 June 1962	14
Denmark	20 May 1965	15
Netherlands	30 March 1967	16
Romania	15 September 1971	17
Germany, DDR	19 November 1974	18
§ Brasil*	16 May 1975	19
Bulgaria	11 September 1978	20
§ Germany, FRG*	5 February 1979	21
§ Uruguay*	11 January 1980	22
Papua New Guinea†	16 March 1981	23
Italy	18 March 1981	24
Peru	10 April 1981	25
Spain	31 March 1982	26
§ China, Peoples' Republic*	8 June 1983	27
§ India*	19 August 1983	28
Hungary	27 January 1984	29
Sweden	24 April 1984	30
Finland	15 May 1984	31
Cuba	16 August 1984	32
Korea, Republic	28 November 1986	33
Greece	8 January 1987	34
Korea, Democratic Peoples Republic	21 January 1987	35

Original signatories (12), which initialled the Treaty on 1 December 1959, are *italicised*; the dates given are those of the deposition of the ratifications of the Treaty.

§ Consultative Parties (18; 12 original signatories and 6 others which achieved this status after becoming actively involved in Antarctic Research).

* These acceding states became Consultative Parties on 29 July 1977 (Poland), 3 March 1981 (Germany, BRD), 12 September 1983 (Brasil and India), and 7 October 1985 (China [Peoples' Republic] and Uruguay).

† Papua New Guinea acceded to the Treaty after becoming independent of Australia.

ACKNOWLEDGEMENTS

Acknowledgements are due to the following for the provision of the pictures on the pages indicated:

Alfred-Wegner-Institut für Polar- und Meeresforschung, Bremerhaven: *Fütterer, 53 (bottom), 54 (top); Hempel, 120; Kohnen, 31*

Antarctic Division, Department of Science and Technology, Tasmania: *19, 22 (left), 76 (bottom), 79 (top)*

British Antarctic Survey, Cambridge: *45 (top and bottom), 77, 80 (top), 99 (top) (with Stanford University); Allan, 109 (bottom), 110 (top), 113; Bonner, 127, 129 (bottom), 130 (top and bottom), 131 (top and bottom); Croxall, 111 (top); Ellis Evans 118 (bottom); Everson, 107; Gilbert, 109 (top); Gipps, 114; Lewis-Smith, 116, 117; Limbert, 11, 75, 78, 79 (bottom), 80 (bottom); McCann 111 (bottom); Rycroft, 97 (bottom); Soar, 76 (top); Swithinbank, 12, 14, 21, 61 (top and bottom), 84 (left and right), 85, 89, Tearle, 119; Wootton, 112*

Camera Press, London: *57*

Comite Nacional de Investigaciones Antarticas, Santiago: *Villanueva, 35, 118*

Cornell University: *100*

Expéditions Polaires Françaises, Paris: *16, 17, 32, 33 (top and bottom), 58, 64, 110 (bottom), 124 (top)*

Hansom, Department of Geography, University of Sheffield: *16*

Hyslop, *10, 48, 56, 59, 68 (top and bottom), 70 (bottom), 71, 86 (top and bottom), 87 (top and bottom), 90, 93, 95, 96, 97 (top), 98, 99 (bottom), 101 (top)*

Izmiran, Moscow, *101 (bottom)*

NASA, Washington: *42, 94*

National Institute of Polar Research, Tokyo: *27 (bottom right); Hirasawa, 52 (top), 102 (bottom); Watanabe, 22 (right), 69 (top and bottom), 70 (top), 88, 92, 104, 106, 124 (bottom), 125 (top and bottom)*

National Science Foundation, Division of Polar Programs, Washington: *20; Hawthorne, 50, 122 (top); Kinne, 32*

New Scientist, London: *24, 25, 116/117*

Norsk Polarinstitutt, Oslo: *53 (top); Eriksson, 15, 46, 52 (bottom), 54 (bottom), 55, 74; Johansen, 73 (top); Orheim, 73 (bottom)*

Royal Society, London: *26, 27 (top)*

Scientific Committee on Antarctic Research, *6*

Scott Polar Research Institute, University of Cambridge: *Drewry, 13, 36, 38 (top), 44, 47, 49, 51, 60, 62 (top and bottom), 63, 65, 66, 67, 82; Antarctic Glaciological and Geophysical Folio, 38, 39 (bottom), 40/41*

Tibbs, Cambridge: *121, 122 (bottom)*

University of Maryland, *Siren, 102 (top)*

INDEX